UNFOLDING OF SYSTEMS OF INDUCTIVE DEFINITIONS

A DISSERTATION
SUBMITTED TO THE DEPARTMENT OF MATHEMATICS
AND THE COMMITTEE OF GRADUATE STUDIES
OF STANFORD UNIVERSITY
IN PARTIAL FULFILLMENT OF THE REQUIREMENTS
FOR THE DEGREE OF
DOCTOR OF PHILOSOPHY

Ulrik Torben Buchholtz
December 2013

Unfolding of Systems of Inductive Definitions
by Ulrik Torben Buchholtz

© 2013 by Ulrik Torben Buchholtz. All Rights Reserved

 This work is licensed under the Creative Commons Attribution 3.0 United States License. To view a copy of this license, visit

http://creativecommons.org/licenses/by/3.0/us/

or send a letter to Creative Commons, 444 Castro Street, Suite 900, Mountain View, California, 94041, USA.

CreateSpace Independent Publishing Platform
North Charleston, SC
ISBN-13: 978-1-4944-0809-1
ISBN-10: 1-4944-0809-0
Library of Congress Control Number: 2013923421

This dissertation is also available online at:
    http://purl.stanford.edu/kg627pm6592

I certify that I have read this dissertation and that, in my opinion, it is fully adequate in scope and quality as a dissertation for the degree of Doctor of Philosophy.

Solomon Feferman, Primary Adviser

I certify that I have read this dissertation and that, in my opinion, it is fully adequate in scope and quality as a dissertation for the degree of Doctor of Philosophy.

Grigori Mints, Co-Adviser

I certify that I have read this dissertation and that, in my opinion, it is fully adequate in scope and quality as a dissertation for the degree of Doctor of Philosophy.

Thomas Strahm

Approved for the Stanford University Committee on Graduate Studies.

Patricia J. Gumport, Vice Provost for Graduate Education

An original signed hard copy of the signature page is on file in University Archives.

ABSTRACT

This thesis is a contribution to Solomon Feferman's Unfolding Program which aims to provide a general method of capturing the operations on individuals and predicates (and the principles governing them) that are implicit in a formal axiomatic system based on open-ended schemata. The thesis in particular studies the unfolding of the classical system for one generalized positive arithmetic inductive definition. The main result is an ordinal analysis of this theory. The resulting ordinal has been known since Heinz Bachmann, and has been studied by Peter Aczel, who felt it should be of proof-theoretic interest. Solomon Feferman conjectured specifically that it should be the strength of the theory under consideration here, and this thesis verifies his conjecture.

The upper bound proceeds via a system of numbers, inductive definitions and ordinals that is analyzed with a combination of operator-controlled derivations and asymmetric interpretation.

The lower bound is established through a well-ordering proof that uses the unfolding machinery to construct hierarchies based on jump operators. This part highlights a new ingredient needed in the unfolding at this level, namely a dependent version of the join operator, producing disjoint unions of predicates indexed by a predicate.

The thesis also includes an appendix detailing the history and motivation of the unfolding program, as well as an appendix describing previous work on the Aczel-Bachmann ordinal.

ACKNOWLEDGEMENTS

This thesis was written under the direction of Professor Solomon Feferman. I would like to thank Professors Solomon Feferman and Grigori Mints for guiding me into the world of proof theory and for all the valuable meetings we had over the last five years.

I am grateful to the following persons for very helpful discussions on my work: Gerhard Jäger, Thomas Strahm, Andreas Weiermann, Stanley S. Wainer, Jeremy Avigad, Rick Statman and Sam Sanders.

I am also grateful to Stanford University and Rejselegat for Matematikere (Danish traveling scholarship for mathematicians) for funding this research.

I thank my wonderful wife, Fortune, for all her support and encouragement during the preparation of this thesis.

TABLE OF CONTENTS

Abstract · v
Acknowledgements · vii
Table of Contents · viii
List of Tables · x
List of Illustrations · xi

1 Unfolding and ID_1 · 1
    1.1 Introduction · 1
        Remark on notation · 2
    1.2 Very brief note on unfoldings · 2
    1.3 The Logic of Partial Terms · 3
        The quantifier-free variant · 4
    1.4 The PCA-version of unfolding · 4
    1.5 The schematic system for one inductive definition · 7

2 Ordinals and notation systems · 9
    2.1 Set-theoretic account · 9
        Ordinal arithmetic · 10, *Cantor normal form* · 11, *Derivatives and the Veblen functions* · 11, *Collapsing functions* · 12
    2.2 The notation system · 13
        Strongly critical components · 15

3 Upper bounds · 17
    3.1 The intermediate system · 18
        The embedding · 20
    3.2 Infinitary languages · 20
        Fragments · 21, *Ranks* · 22, *Characteristic sequences* · 22
    3.3 The Strategy · 23
        Acceptable operators · 23
    3.4 The restricted infinitary calculus · 24
    3.5 The derived infinitary calculus · 29
    3.6 Connecting the systems · 31

- 3.7 Reduction · 33
- 4 Lower bounds · 35
  - 4.1 Preliminaries · 35
  - 4.2 Outline · 35
  - 4.3 The accessible part · 36
  - 4.4 Stepping up · 39
  - 4.5 Condensation · 45
- 5 Conclusion · 47
  - 5.1 Further work · 47
    Restricted join · 47 , *The unfolding of* ID_ν · 48 , *Connections with iterated fixed points* · 48 , *Parametrized arithmetical fixed points* · 49 , *Intuitionistic theories and type theories* · 49 , *Unfolding of set theory* · 50
- A History of Unfoldings · 51
  - A.1 Motivation of the unfolding program · 51
  - A.2 Early forerunners of unfoldings · 52
    Reflection principles · 53 , *Autonomous progressions* · 54
  - A.3 Schematic theories · 55
  - A.4 Reflective closure · 56
    Relations to paradoxes · 58 , *Other truth theories* · 59
  - A.5 Theories of operations · 59
    Another motivation for operations · 60
  - A.6 Unfoldings · 61
    Previous results concerning unfoldings · 63 , *Criticisms of foundations* · 64
  - A.7 Relations to Frege Structures · 64
    Unfolding in Frege Structures · 65 , *Undefinability results* · 65
  - A.8 Relations to Explicit Mathematics · 65
- B Ordinal iteration and Aczel's ordinal · 67
  - B.1 The G functionals · 69
  - B.2 The H functionals · 69
  - B.3 Comparison of ordinals · 70
- Bibliography · 75
- Symbol Index · 82
- Index · 83

LIST OF TABLES

5.1 Strengths of theories; the ones below the line are conjectured. · 48

LIST OF ILLUSTRATIONS

3.1 Relation between infinitary systems for the upper bound · 30
A.1 Kleene's strong three-valued logic · 58

Chapter 1

UNFOLDING AND ID$_1$

1.1   INTRODUCTION

This thesis is concerned with the unfoldings of schematic systems of inductive definitions. The bulk of the following material is devoted to proving the main result, which is the calculation of the proof-theoretic ordinal of the unfolding of the schematic theory of one generalized inductive definition, ID$_1$:

THEOREM 1.1.1 (Main Theorem). $|\mathcal{U}(\text{ID}_1)| = \psi(\Gamma_{\Omega+1})$.

This solves a problem posed by Feferman, which in turn stems from a "feeling" of Aczel and a loose conjecture of Miller. The ordinal of the theorem is equal to $H(1)$ from BACHMANN (1950) (cf. Appendix B.3) and had turned up in investigations on iterations of functionals of ordinals by ACZEL (1972). Therefore, Aczel felt this ordinal should be of proof-theoretic interest:

> By analogy with the proof theoretic significance of ε_0, Γ_0 and $\varphi_{\varepsilon_{\Omega+1}}(0)$ we feel that $H(1)$ may also be of similar interest and conjecture that it may be the proof theoretic ordinal of a suitable system of ramified analysis.
>
> ACZEL 1972, p. 36

Later, Miller conjectured

> ... $H(1) = (1; \Omega)$ to be the proof-theoretic ordinal of ID$_1^*$ which is related to ID$_1$ as predicative analysis ID$_0^*$ is to first-order arithmetic ID$_0$, i.e., predicative construction given the inductively defined set.   MILLER 1976, p. 451

But Miller did not specify a precise system ID$_1^*$. Feferman's general notion of unfolding gives us such a system, $\mathcal{U}(\text{ID}_1)$, and indeed we may compare our result with that of FEFERMAN AND STRAHM (2000), according to which $|\mathcal{U}(\text{NFA})| = \Gamma_0$, where NFA is a basic schematic system of Non-Finitist Arithmetic. Recall the theorem of

Gerber (1970) and Howard (1972) that $|ID_1| = \psi(\varepsilon_{\Omega+1})$.[1] Thus the ordinal of Peano Arithmetic, ε_0 is to Γ_0 as $\psi(\varepsilon_{\Omega+1})$ is to $\psi(\Gamma_{\Omega+1})$.

The remainder of the current chapter provides the exact definitions of the unfolding systems and of the theory of schematic ID_1.

Chapter 2 introduces the ordinal notation systems that we shall use to analyze proof-theoretically the various unfolding systems.

Chapter 3 goes through the upper bounds results.

Chapter 4 details the lower bounds via explicit well-ordering proofs.

Chapter 5 summarizes our contribution and discusses possibilities for further work.

I have put an exposition of the history and motivation of the unfolding program in Appendix A.

The ordinal occurring in the main theorem, $\psi(\Gamma_{\Omega+1})$ is also known as Aczel's ordinal. I describe why, and where else in the literature it has appeared, in Appendix B.

* * *

Before we get started, I should mention that most the discussion will be focused on theories formulated in first-order classical logic. There will be connections to constructivism here and there, e.g., when we discuss Frege structures and Explicit Mathematics, but we will leave a thorough treatment of "unfolding" of theories in intuitionistic frameworks for another occasion.

Remark on notation

Throughout the thesis, I will use the dot-notation for quantifiers to reduce the amount of parentheses. So I will write $\forall x.\, A(x) \vee B(x)$ instead of $(\forall x)(A(x) \vee B(x))$, for example. The dot represents an opening bracket with a corresponding closing bracket as far to the right as syntactically possible.

There is an index of symbols at the end of the thesis to serve as a guide to specific notations introduced in the work.

I.2 VERY BRIEF NOTE ON UNFOLDINGS

The idea of the unfolding of an open-ended schematic axiom system was introduced in Feferman (1996). It is explained very briefly via the following quotation:

---

[1] Using the Buchholz ψ-function instead of Bachmann's $\varphi_{\varepsilon_{\Omega+1}}(0)$.

> Given a schematic system S, which operations and predicates, and which principles concerning them, ought to be accepted if one has accepted S?
>
> FEFERMAN (1996, §3)

The operations with which one is concerned include those that apply to individuals and those that apply to predicates (such as logical connectives). For the full background to the notion of unfolding and earlier results by FEFERMAN AND STRAHM (2000, 2010) see Appendix A.

As is described in more detail in that appendix, the unfoldings rely on either a partial combinatory algebra structure or the theory of generalized recursive definability in order to capture the operations involved. Either way relies in turn on the logic of partial terms which we recall in Section 1.3. After that, Section 1.4 describes the version using a partial combinatory algebra, as that is the more modern and smoother formulation.[2]

Section 1.5 describes precisely the schematic system for one arithmetical generalized inductive definition. The unfolding of this system is the main object of interest in the rest of the work.

1.3 THE LOGIC OF PARTIAL TERMS

In this section we will review the usual Logic of Partial Terms (BEESON 1986) as well as a quantifier-free version. The atomic formulæ include those of the form $t\downarrow$, for each term t, expressing that t is defined.

We define
$$s \simeq t :\Leftrightarrow t\downarrow \vee s\downarrow \to s = t.$$
Then we have the following rules and axioms, starting with the modified quantifier

---
[2] The original version of unfolding in terms of recursive definability with least fixed-point recursion is described in Appendix A.6.

rules

(Q1)	$\dfrac{B \to A}{B \to \forall x.\, A}$	if x is not free in B,
(Q2)	$\dfrac{A \to B}{(\exists x.\, A) \to B}$	if x is not free in B,
(Q3)	$(\forall x.\, A) \wedge t{\downarrow} \to A[x := t]$,	
(Q4)	$A[x := t] \wedge t{\downarrow} \to \exists x.\, A$,	
(E1)	$x = x \wedge (x = y \to y = x)$,	
(E2)	$t \simeq s \wedge A(t) \to A(s)$,	
(E3)	$t = s \to t{\downarrow} \wedge s{\downarrow}$,	
(S1)	$R(t_1, \ldots, t_n) \to t_1{\downarrow} \wedge \cdots \wedge t_n{\downarrow}$,	
(S2)	$c{\downarrow}$ for constants c,	
(S3)	$x{\downarrow}$ for variables x.	

The quantifier-free variant

For the quantifier-free variant of LPT, we remove (Q1) and (Q2) and replace (Q3) and (Q4) by

(Q5) $\qquad\qquad\qquad \dfrac{A \quad t{\downarrow}}{A[x := t]}\,.$

I.4 THE PCA-VERSION OF UNFOLDING

One of the operations on predicates used in the definition of unfolding in FEFERMAN (1996) and FEFERMAN AND STRAHM (2000) was that of countable join (expressed as join over the universe). In comparison with these, we add here a *dependent version* of the join operation. In a Frege structure the analogous operation is *sequential implication* or *short circuit conjunction*. This seems to be necessary for the lower bound proof, and does not affect the upper bound proof.[3] The dependent join operation is very natural, so we have no qualms about adding it.

The operational unfolding is based on the notion of a partial combinatory algebra:

---

[3] It is currently an open problem whether the system with join only over ω is as strong as this one.

DEFINITION 1.4.1. A *partial combinatory algebra* (PCA) is a set A equipped with a partial "application" operation $\cdot : A \times A \rightharpoonup A$ and elements $k, s \in A$ such that for all $x, y, z \in A$:

$$k \cdot x \cdot y = x, \quad s \cdot x \cdot y \downarrow, \quad s \cdot x \cdot y \cdot z \simeq (x \cdot z) \cdot (y \cdot z).$$

Here, $t \simeq s$ is the relation that if either of the closed terms t and s denotes an element of A, then so does the other and the denoted elements are equal.

The leading example is Kleene's first algebra K_1, consisting of the natural numbers with *Kleene application*, $m \cdot n \simeq \{m\}(n)$; Kleene's second algebra, K_2, gives a PCA over $\mathbb{N}^\mathbb{N}$ (cf. KLEENE 1945; KLEENE AND VESLEY 1965). In the following we shall omit the · symbol and hence denote application simply by juxtaposition.

Another example is given by the term model for the untyped λ-calculus.

Suppose S is a schematic first-order system, possibly formulated with many sorts. In the PCA-based version of unfolding, we define the operational unfolding, $\mathcal{U}_0(S)$, in a single-sorted language of operations into which we embed the sorts of S by means of relativizing predicates, τ. We add constants to denote the function symbols of S, and new constants k and s (combinators), p, p_0 and p_1 (pairing and projection), d, tt and ff (definition by cases, true and false), e (equality), and the binary function symbol · (application). The relation symbols are the unary symbol $\downarrow$ (defined), the binary symbol = (equality), as well as unary predicates for the sorts of S, and the actual predicates of S itself. We write (a, b) for $p\, a\, b$.

The axioms of $\mathcal{U}_0(S)$, the operational unfolding, are
1. Embedding of S:
    a) The relativization of the axioms of S to the appropriate sorts, and axioms declaring that the function symbols of S are total on the appropriate sorts.
2. Partial combinatory algebra (PCA) axioms with pairing and definition by cases:
    a) $k\, a\, b = a$.
    b) $s\, a\, b \downarrow \wedge s\, a\, b\, c \simeq a\, c\, (b\, c)$.
    c) $p_0(a, b) = a \wedge p_1(a, b) = b$.
    d) $d\, a\, b\, \text{tt} = a \wedge d\, a\, b\, \text{ff} = b$.
3. Equality on the sorts τ for which equality should be operationally decidable:
    a) $\forall x, y \in \tau.\, e\, x\, y = \text{tt} \vee e\, x\, y = \text{ff}$.
    b) $\forall x, y \in \tau.\, e\, x\, y = \text{tt} \leftrightarrow x = y$.

In addition, $\mathcal{U}_0(S)$ includes the unrestricted substitution rule.

Abstraction terms $\lambda x.\, t$ can be defined as usual, and from the PCA axioms we

can show in $\mathcal{U}_0(S)$:
1. $(\lambda x.t)\!\downarrow \wedge (\lambda x.t)\,x \simeq t$
2. $s\!\downarrow \,\to\, (\lambda x.t)\,s \simeq t[x := s]$

Note that we use the notation of the λ-calculus even though the conversion relation is not exactly the same (in particular it does not validate the (ξ)-rule of the λ-calculus).

The PCA axioms allow us to introduce a fixed point operator, but we cannot prove that it produces *least* fixed points.

THEOREM 1.4.2 (Fixed point). *There is a closed term* fix *of* $\mathcal{U}_0(S)$ *so that*

$$\mathcal{U}_0(S) \vdash \text{fix}\, f\!\downarrow \wedge \text{fix}\, f\, x \simeq f\,(\text{fix}\,f)\,x.$$

The language of the full unfolding $\mathcal{U}(S)$ in this formulation extends the language by additional constants for the sorts, $\dot{\tau}$, and predicates, p, of S, eq (equality), pr_U (free predicate U), inv (inverse image), conj (conjunction), neg (negation), un (universal quantification over the natural numbers), join (join, that is, disjoint union). In addition, we add the unary relation symbol Π (predicates) and the binary relation symbol $\in$ (predication). The axioms of $\mathcal{U}(S)$ extend the ones of $\mathcal{U}_0(S)$ by

4. Basic axioms about predicates:
   a) $\Pi(\dot{\tau}) \wedge \forall x.\, x \in \dot{\tau} \leftrightarrow \tau(x)$.
   b) $\Pi(p) \wedge \forall x.\, x \in p \leftrightarrow P(x)$.
   c) $\Pi(\text{eq}) \wedge \forall x.\, x \in \text{eq} \leftrightarrow \exists y.\, x = (y, y)$.
   d) $\Pi(\text{pr}_U) \wedge \forall x.\, x \in \text{pr}_U \leftrightarrow U(x)$.
   e) $\Pi(a) \to \Pi(\text{inv}(a, f)) \wedge \forall x.\, x \in \text{inv}(a, f) \leftrightarrow f\, x \in a$.
   f) $\Pi(a) \wedge \Pi(b) \to \Pi(\text{conj}(a, b)) \wedge \forall x.\, x \in \text{conj}(a, b) \leftrightarrow x \in a \wedge x \in b$.
   g) $\Pi(a) \to \Pi(\text{neg}\, a) \wedge \forall x.\, x \in \text{neg}(a) \leftrightarrow \neg(x \in a)$.
   h) $\Pi(a) \to \Pi(\text{un}\, a) \wedge \forall x.\, x \in \text{un}(a) \leftrightarrow \forall y \in \mathbb{N}.\, (x, y) \in a$.

5. The dependent join axiom:

$$\Pi(a) \wedge (\forall y \in a.\, \Pi(f\, y)) \to \Pi(\text{join}(f, a))$$
$$\wedge \forall x.\, x \in \text{join}(f, a) \leftrightarrow \exists y, z.\, x = (y, z) \wedge y \in a \wedge z \in f(y).$$

Finally, $\mathcal{U}(S)$ contains the restricted substitution rule:

$$\frac{A(U)}{A(\{x \mid B(x)\})}\ (\text{SUBST})$$

Here, B is any formula, and A is any formula in the language of $\mathcal{U}_0(S)$.

I.5 THE SCHEMATIC SYSTEM FOR ONE INDUCTIVE DEFINITION

The system we'll be unfolding is a schematic version of ID_1, the system that formalizes one arithmetical inductive definition. The language of ID_1, $\mathcal{L}^1$, is that of (single-sorted) first-order arithmetic, PA, (with a free predicative variable U) augmented by a predicate symbol $I_\mathcal{O}$ for an arithmetical positive operator form $\mathcal{O}(X,x)$ that does not contain U (we use the letter $\mathcal{O}$ because the universal case is that of Kleene's $\mathcal{O}$ describing the second constructive number class).

The non-logical axioms and rules of our schematic version of ID_1 are:

Number-theoretic axioms: The axioms of PA with exception of the induction scheme.

Schematic induction axiom on the natural numbers:

$$U(0) \wedge (\forall x. U(x) \to U(x')) \to \forall x. U(x).$$

Schematic inductive definition axioms:

$$\forall x. \mathcal{O}(I_\mathcal{O}, x) \to I_\mathcal{O}(x) \quad \text{and}$$
$$(\forall x. \mathcal{O}(U, x) \to U(x)) \to \forall x. I_\mathcal{O}(x) \to U(x).$$

Substitution rule: For A and $B(x)$ formulæ of ID_1:

$$\frac{A(U)}{A(\{x \mid B(x)\})} \text{ (SUBST)}$$

Chapter 2

ORDINALS AND NOTATION SYSTEMS

This chapter introduces the necessary preliminaries on ordinals and notation systems that we shall use to analyze the unfoldings of schematic theories for generalized inductive definitions in the following chapters.

The ordinal notation system and its arithmetical coding that is used here is taken from POHLERS (2009, Chapter 9). We denote codes of ordinals by lowercase italic Greek letters and overload the $<$ relation (and the arithmetical operations) to write $\alpha < \beta$ for the ordering relation of the notation system.

2.1 SET-THEORETIC ACCOUNT

This section describes the preliminaries of the usual set-theoretic account of ordinals in order to describe the ordinals and associated notation systems needed to understand our proof-theoretic results in the following chapters.

First we explain how to understand the ordinal functions and the collapsing operations in terms of the usual set-theoretical account in terms of ordinals as hereditarily transitive sets. In ZFC this entails that ordinals are well-founded with the $\in$-relation, and each ordinal is equal to the set of its predecessors.

Every ordinal falls into one of the following classes:
- the least ordinal, 0;
- successor ordinals, i.e., ordinals of the form $\alpha' = \alpha \cup \{\alpha\}$;
- limit ordinals. The class of limit ordinals is denoted Lim.

Recall also that a *regular ordinal* is an ordinal $\kappa \in \text{Lim}$ such that whenever $M \subseteq \kappa$ (the cardinality of M, as an initial ordinal, is less than κ) has $|M| < \kappa$, we have $M \subseteq \alpha < \kappa$ for some α (M is *bounded* in κ). The class of regular ordinals larger than ω is denoted Reg, and it follows in ZFC that Reg is unbounded in the class of all ordinals, On.

Every subclass $M \subseteq \text{On}$ is well-ordered with an order-type $\leq$ On. If M is bounded in On, its order-type is an ordinal $\alpha \in \text{On}$ and we have an enumeration

9

function $\text{en}_M: \alpha \to M$. If M is unbounded in On, it has the order-type of On itself and we have an enumeration function $\text{en}_M: \text{On} \to M$.

Let $\lambda\xi.\,\Omega_\xi$ be the enumeration function for $\{0\} \cup \text{Reg}$. We set $\Omega := \Omega_1$.

A class $M \subseteq \text{On}$ is called *closed*, if M is closed in the order-topology, which is the case precisely when suprema of elements in M themselves are in M.

Let $\kappa \in \text{Reg}$. An ordinal function $f: \kappa \to \kappa$ (or $f: \text{On} \to \text{On}$) is *continuous* when it preserves suprema. And f is called *normal* when it is strictly increasing and continuous. A normal function satisfies $f(\alpha) \geq \alpha$ for all α.

A class $M \subseteq \kappa$ is called κ-*club* if and only if it is *closed* and *unbounded* in κ. A fundamental fact is that M is club if and only its enumeration function en_M is normal.

Ordinal arithmetic

We can extend the usual recursive definition of addition, multiplication and exponentiation to ordinal numbers so that they are continuous in their second argument. First, addition:

$$\alpha + 0 := \alpha$$
$$\alpha + (\beta + 1) := (\alpha + \beta) + 1$$
$$\alpha + \lambda := \sup\{\alpha + \xi \mid \xi < \lambda\} \quad \text{for } \lambda \in \text{Lim}.$$

Secondly, multiplication:

$$\alpha \cdot 0 := 0$$
$$\alpha \cdot (\beta + 1) := (\alpha \cdot \beta) + \alpha$$
$$\alpha \cdot \lambda := \sup\{\alpha \cdot \xi \mid \xi < \lambda\} \quad \text{for } \lambda \in \text{Lim}.$$

And finally, exponentiation:

$$\alpha^0 := 1$$
$$\alpha^{\beta+1} := \alpha^\beta \cdot \alpha$$
$$\alpha^\lambda := \sup\{\alpha^\xi \mid \xi < \lambda\} \quad \text{for } \lambda \in \text{Lim}.$$

It is easy to see that the class $\{\xi \mid \xi \geq \alpha\}$ is club in all regular $\kappa > \alpha$, and $\lambda\xi.\,\alpha + \xi$ is its enumerating function which is hence normal.

We let $\mathbb{H}$ be the class of *additively indecomposable* ordinals (so $\alpha \in \mathbb{H}$ if and only if for all $\xi, \eta < \alpha$, $\xi + \eta < \alpha$). Then $\mathbb{H}$ is κ-club for all $\kappa \in \mathrm{Reg}$ and its enumerating function is $\lambda \xi. \omega^\xi$.

Cantor normal form

By induction on $\alpha > 0$, we prove that there are uniquely determined ordinals $\{\alpha_1, \ldots, \alpha_n\} \subseteq \mathbb{H}$ such that

$$\alpha = \alpha_1 + \cdots + \alpha_n \quad \text{and} \quad \alpha_1 \geq \cdots \geq \alpha_n.$$

This is called the *additive normal form* of α, and $\{\alpha_1, \ldots, \alpha_n\}$ is called the set of *additive components* of α.

Using the enumerating function for $\mathbb{H}$ we obtain the *Cantor normal form* for $\alpha > 0$:

$$\alpha =_{\mathrm{NF}} \omega^{\xi_1} + \cdots + \omega^{\xi_n} \quad \text{with} \quad \xi_1 \geq \cdots \geq \xi_n.$$

Any ordinal less than the first fixed point ε_0 of $\lambda \xi. \omega^\xi$ has a unique representation in hereditary Cantor normal form starting from 0.

Derivatives and the Veblen functions

For a class $M \subseteq \kappa$ we define its derivative as the class of the fixed points of its enumerating function:

$$M' := \{\xi < \kappa \mid \mathrm{en}_M(\xi) = \xi\}$$

The *derivative* f' of a function f is defined by $f' := \mathrm{en}_{\mathrm{Fix}(f)}$ where

$$\mathrm{Fix}(f) := \{\xi < \kappa \mid f(\xi) = \xi\}.$$

It is easily shown that if M is κ-club, then so is M'. Hence, if f is κ-normal, then so is f'.

If $\{M_\xi\}$ is a collection of less than κ many κ-club classes, then $\bigcap_\xi M_\xi$ is again κ-club.

These observations allows us to iterate the derivation process transfinitely to obtain a hierarchy of club classes. If we start this process with $\mathbb{H}$ we obtain

$$\begin{aligned}
\operatorname{Cr}(0) &:= \mathbb{H} \\
\operatorname{Cr}(\alpha + 1) &:= \operatorname{Cr}(\alpha)' \\
\operatorname{Cr}(\lambda) &:= \bigcap_{\xi < \lambda} \operatorname{Cr}(\xi) \quad \text{for } \lambda \in \operatorname{Lim}.
\end{aligned}$$

Then we put $\varphi_\alpha := \operatorname{en}_{\operatorname{Cr}(\alpha)}$ to obtain the usual *Veblen hierarchy*.

Ordinals closed under the binary Veblen function, $\lambda\alpha, \beta. \varphi_\alpha(\beta)$, are called *strongly critical ordinals*. Their class, SC, is κ-club for $\kappa > \omega$, and we let $\lambda\xi. \Gamma_\xi$ be their enumerating function. The first strongly critical ordinal, Γ_0, is also known as the Feferman-Schütte ordinal, and is well-known from the analysis of predicativity given the natural numbers.

If $\alpha \in \mathbb{H} \setminus \operatorname{SC}$, then α can be written in normal form

$$\alpha =_{\operatorname{NF}} \varphi_{\alpha_1}(\alpha_2), \qquad \text{with } \alpha_1, \alpha_2 < \alpha.$$

Collapsing functions

We recall here the ψ-family of functions, first introduced by Buchholz in an unpublished manuscript (1981) (cf. BUCHHOLZ (1986)).

DEFINITION 2.1.1. By recursion on α we define simultaneously

$$\begin{aligned}
\operatorname{Cl}(\alpha, \beta) &:= \text{the least set containing } \beta \cup \{0, \Omega\} \\
&\quad \text{and closed under } +, \text{ the Veblen function } \lambda\xi\eta. \varphi_\xi(\eta), \\
&\quad \text{and the restricted function } \psi \upharpoonright \alpha := \lambda\xi < \alpha. \psi(\xi), \\
\psi(\alpha) &:= \min\{\beta \mid \operatorname{Cl}(\alpha, \beta) \cap \Omega \subseteq \beta\}.
\end{aligned}$$

We record a few standard facts about these closure sets and the corresponding collapsing function:

LEMMA 2.1.2. *Define* $\operatorname{Cl}_\Omega(\alpha) := \operatorname{Cl}(\alpha, \psi(\alpha))$. *We have*
1. *If* $\beta < \Omega$, *then* $\operatorname{Cl}(\alpha, \beta)$ *is countable.*
2. $\operatorname{Cl}_\Omega(\alpha) \cap \Omega = \psi(\alpha)$.
3. $\psi(\alpha) < \Omega$ *and* $\psi(\alpha) \notin \operatorname{Cl}_\Omega(\alpha)$.
4. $\psi(\alpha) \in \operatorname{SC}$.

5. If $\alpha \leq \beta$, then $\psi(\alpha) \leq \psi(\beta)$ and $\mathrm{Cl}_\Omega(\alpha) \subseteq \mathrm{Cl}_\Omega(\beta)$.
6. If $\alpha < \beta$ and $\alpha \in \mathrm{Cl}_\Omega(\beta)$, then $\psi(\alpha) < \psi(\beta)$.
7. ψ is continuous.

We refer to POHLERS (1989) for more details and proofs.

Informally, we think of $\mathrm{Cl}(\alpha, 0)$ as the α-th iterated Skolem hull of $\{0, \Omega\}$. Then $\mathrm{Cl}(\alpha, 0) = \mathrm{Cl}_\Omega(\alpha)$, so we think of the ordinal $\psi(\alpha) = \mathrm{Cl}_\Omega(\alpha) \cap \Omega$ as the α-th iterated *collapse* of Ω.

2.2 THE NOTATION SYSTEM

For reference, we give here the definition of the notation system. It is built from terms representing 0 and Ω, and function symbols representing n-ary addition, $\bar{\varphi}$ (fixed-point-free version of the binary Veblen function), and the collapsing function ψ. The $\bar{\varphi}$-function is defined by:

$$\bar{\varphi}_\alpha(\beta) := \begin{cases} \varphi_\alpha(\beta + 1), & \text{if } \beta = \gamma + n \text{ with } \varphi_\alpha(\gamma) = \gamma \text{ or } \varphi_\alpha(\gamma) = \alpha, \\ \varphi_\alpha(\beta), & \text{otherwise.} \end{cases}$$

We define sets $\mathrm{SC} \subseteq \mathrm{H} \subseteq \mathrm{OT}$ of ordinal notations (with ordinals in SC denoting strongly critical ordinals, and ordinals in H denoting additively principal ordinals) together with finite sets of subterms $\mathrm{K}(\alpha) \subseteq \mathrm{OT}$ for each notation α, and a relation $< \subseteq \mathrm{OT} \times \mathrm{OT}$. These are defined by the following mutually recursive clauses:

- Definition of $\mathrm{SC} \subseteq \mathrm{H} \subseteq \mathrm{OT}$:
    - $\langle 0 \rangle \in \mathrm{OT}$ (denoting 0),
    - $\langle 1 \rangle \in \mathrm{SC}$ (denoting Ω),
    - if $n > 1$, $\alpha_1, \ldots, \alpha_n \in \mathrm{H}$ and $\alpha_1 \geq \cdots \geq \alpha_n$, then $\langle 2, \alpha_1, \ldots \alpha_n \rangle \in \mathrm{OT}$ (denoting $\alpha_1 + \cdots + \alpha_n$),
    - $\alpha, \beta \in \mathrm{OT}$, then $\langle 3, \alpha, \beta \rangle \in \mathrm{H}$ (denoting $\bar{\varphi}_\alpha \beta$).
    - if $\alpha \in \mathrm{OT}$ and $\mathrm{K}(\alpha) \subseteq \alpha$, then $\langle 4, \alpha \rangle \in \mathrm{SC}$ (denoting $\psi(\alpha)$).

- Definition of K(α):

$$K(0) := \emptyset,$$
$$K(\Omega) := \emptyset,$$
$$K(\alpha_1 + \cdots + \alpha_n) := K(\alpha_1) \cup \cdots \cup K(\alpha_n),$$
$$K(\bar{\varphi}_\alpha \beta) := K(\alpha) \cup K(\beta),$$
$$K(\psi(\alpha)) := \{\alpha\} \cup K(\alpha).$$

- For $\alpha, \beta \in \mathrm{OT}$, put $\alpha < \beta$ if one of the following conditions obtains:
  - $\alpha = 0$ and $\beta \neq 0$,
  - $\alpha = \alpha_1 + \cdots + \alpha_m$, $\beta = \beta_1 + \cdots + \beta_n$, and either
    * $m \geq n$ and $\exists i \leq n. \alpha_i < \beta_i \wedge \forall j < i. \alpha_j = \beta_j$, or
    * $m < n$ and $\forall i \leq m. \alpha_i = \beta_i$.
  - $\alpha = \alpha_1 + \cdots + \alpha_n$, $\beta \in \mathrm{H}$, and $\alpha_1 < \beta$.
  - $\alpha \in \mathrm{H}$, $\beta = \beta_1 + \cdots + \beta_n$, and $\alpha \leq \beta_1$.
  - $\alpha = \bar{\varphi}_{\alpha_1}\alpha_2$, $\beta = \bar{\varphi}_{\beta_1}\beta_2$ and one of the following obtains
    * $\alpha_1 < \beta_1$ and $\alpha_2 < \beta$.
    * $\alpha_1 = \beta_1$ and $\alpha_2 < \beta_2$.
    * $\beta_1 > \alpha_1$ and $\beta_2 \leq \alpha$.
  - $\alpha = \bar{\varphi}_{\alpha_1}\alpha_2$, $\beta \in \mathrm{SC}$, and $\alpha_1, \alpha_2 < \beta$.
  - $\alpha \in \mathrm{SC}$, $\beta = \bar{\varphi}_{\beta_1}\beta_2$, and $\alpha \leq \beta_1$ or $\alpha \leq \beta_2$.
  - $\alpha = \psi(\alpha_1)$, $\beta = \psi(\beta_1)$ and $\alpha_1 < \beta_1$.
  - $\alpha = \psi(\alpha_1)$ and $\beta = \Omega$.

Note that the coefficient sets K make sure we build terms of normal form, where we define
$$\alpha =_{\mathrm{NF}} \psi(\xi) :\Leftrightarrow \alpha = \psi(\xi) \wedge \xi \in \mathrm{Cl}_\Omega(\xi).$$

Strongly critical components

DEFINITION 2.2.1. Given a notation α, we define its set of *strongly critical components*, $SC(\alpha)$, as follows:

$$SC(0) := \emptyset,$$
$$SC(\alpha) := \{\alpha\}, \quad \text{if } \alpha \in SC,$$
$$SC(\alpha_1 + \cdots + \alpha_n) := SC(\alpha_1) \cup \cdots \cup SC(\alpha_n),$$
$$SC(\bar{\varphi}_\alpha \beta) := SC(\alpha) \cup SC(\beta).$$

LEMMA 2.2.2. *We have:*
- *For all α, $SC(\alpha) \subseteq \alpha + 1$.*
- *If $\alpha \leq \beta$ and $\alpha \in SC$, then $\alpha \leq \mu$ for some $\mu \in SC(\beta)$.*

LEMMA 2.2.3. *There are primitive recursive operations, $+$ and φ, such that if $\alpha, \beta \in OT$, then $\alpha + \beta, \varphi_\alpha(\beta) \in OT$, and these denote respectively the result of applying addition and the φ-function to the denotations of α and β. Furthermore, $SC(\alpha + \beta) \cup SC(\varphi_\alpha(\beta)) \subseteq SC(\alpha) \cup SC(\beta)$.*

Proof. The operations can be defined by recursion on the structure of α and β (since we can code primitive recursive predicates for being ξ-critical). Inspection of the resulting terms then establishes the second claim. □

LEMMA 2.2.4. *There is a primitive recursive operation that for each $\alpha \in OT$ determines a list $\alpha_1, \ldots, \alpha_k \in OT$ giving a Cantor Normal Form*

$$\alpha = \omega^{\alpha_1} + \cdots + \omega^{\alpha_n}, \qquad \alpha_1 \geq \cdots \geq \alpha_n.$$

Furthermore, $SC(\alpha_i) \subseteq SC(\alpha)$ for each i.

Chapter 3

UPPER BOUNDS

In this chapter we will discuss the upper bound part of the main result, that is, we will show that the proof-theoretic ordinal of $\mathcal{U}(\mathrm{ID}_1)$ is at most $\psi(\Gamma_{\Omega+1})$. The proof uses an intermediate system, $(\mathrm{ID}_1)_{\mathrm{On}}^+ + (\mathrm{Subst})$, that is then interpreted in a pair of infinitary systems for languages describing stages of inductive operators.

The intermediate system is exactly analogous to the one used by FEFERMAN AND STRAHM (2000) in their upper bound proof for the strength of $\mathcal{U}(\mathrm{NFA})$. The strength of that system, $\mathrm{PA}_\Omega^+ + (\mathrm{Subst})$, was determined by STRAHM (2000) using a triple of semiformal systems for numbers and ordinals. We shall use a similar setup, but there are a number of complications in the step to ID_1:

- For derivations of formulæ in (the translation of) of the language of $\mathcal{U}_0(\mathrm{ID}_1)$ we need to eliminate all cuts on formulæ from the full language. However, complete cut-elimination for ID_1 is typically not obtainable, so we need a separate language to analyze the inductive predicate $I_\emptyset$.
- Because of the lack of complete cut-elimination, substitution is harder to deal with.
- Because we are at the level of impredicative proof theory, we need some device to handle collapse. Here we shall use operator-controlled derivations in the style of BUCHHOLZ (1992).

First, in Section 3.1 we discuss the intermediate system, $(\mathrm{ID}_1)_{\mathrm{On}}^+ + (\mathrm{Subst})$ and how it embeds the unfolding.

Next, to find the upper bound for the intermediate we shall set up the necessary languages in Section 3.2. Section 3.3 discusses the strategy in more detail. In Section 3.4 we set up an infinitary calculus for treating both the inductive predicate of ID_1 and the stages of the operators of $(\mathrm{ID}_1)_{\mathrm{On}}^+ + (\mathrm{Subst})$. The axioms of the intermediate system itself are handled in a derived system which is described in Section 3.5. These systems are then compared in Section 3.6, so we in Section 3.7 can finish the reduction.

3.1 THE INTERMEDIATE SYSTEM

We shall use an exact analogue of the intermediate system of FEFERMAN AND STRAHM (2000), $PA_\Omega^+ + (Subst)$, which we shall call $(ID_1)_{On}^+ + (Subst)$.[1] This system interprets the full unfolding of schematic ID_1. The only difference is the interpretation of the dependent join operation. But since the predicates $\Pi(a)$, $x \in a$ and $x \bar{\in} a$ are defined by a simultaneous inductive definition, the extra dependency caused by the dependent join operation does not spoil anything.

The theory $(ID_1)_{On}^+ + (Subst)$ is formulated in the language $\mathcal{L}_{On}^1$, which is obtained from the language of ID_1, $\mathcal{L}^1$, by adding a new sort for ordinal variables (with $<$ and $=$ relations), and an $(n+1)$-ary predicate symbol $P_\mathfrak{A}$ for each inductive operator form $\mathfrak{A}(X, \vec{x})$ over ID_1 (that is, $\mathfrak{A}$ is an $\mathcal{L}^1$-formula so it can contain U and both positive and negative occurrences of $I_\mathcal{O}$, but of course only positive occurrences of the fresh n-ary predicate variable X).

The formulæ of $\mathcal{L}_{On}^1$ are built from these literals using conjunction, disjunction, quantification over the numbers, bounded quantification over the ordinals, and unbounded quantification over the ordinals. Negation is defined via de Morgan's laws and double negation elimination. Of particular interest are the natural fragments of Δ-, Σ- and Π-formulæ that are built using no unbounded quantifiers, no unbounded universal quantifiers, and no unbounded existential quantifiers, respectively. For a formula A and an ordinal variable ξ, A^ξ denotes the formula that is obtained from A by bounding all unbounded ordinal quantifiers by ξ.

As a matter of notation we write $P_\mathfrak{A}^\alpha(\vec{x})$ instead of $P_\mathfrak{A}(\alpha, \vec{x})$, and we put $P_\mathfrak{A}^{<\alpha}(\vec{x}) := \exists \beta < \alpha. P_\mathfrak{A}^\beta(\vec{x})$.

Note that we use two notational devices to distinguish the inductive predicates of ID_1 from the stages of the operators in $\mathcal{L}_{On}$: $I_\mathcal{O}$ is the least fixed point of an arithmetical operator, while the $P_\mathfrak{A}^\alpha$ form the stages of an inductive definition on top of that.

$(ID_1)_{On}^+ + (Subst)$ is axiomatized as follows:

Number-theoretic axioms: The axioms of PA with exception of the induction scheme.
Schematic induction axiom on the natural numbers:

$$U(0) \wedge (\forall x. U(x) \to U(x')) \to \forall x. U(x).$$

---

[1] Since Ω for us denotes the first uncountable regular cardinal, we use On to indicate the formal system with ordinals.

Schematic inductive definition axioms:

$$\forall x. \mathcal{O}(I_\mathcal{O}, x) \to I_\mathcal{O}(x) \quad \text{and}$$
$$(\forall x. \mathcal{O}(U, x) \to U(x)) \to \forall x. I_\mathcal{O}(x) \to U(x).$$

Inductive operator axioms: For any inductive operator $\mathfrak{A}(X, \vec{x})$:

$$P_\mathfrak{A}^\sigma(\vec{x}) \leftrightarrow \mathfrak{A}(P_\mathfrak{A}^{<\sigma}, \vec{x}).$$

Linearity axioms:

$$\sigma \not< \sigma \wedge (\sigma < \tau \wedge \tau < \eta \to \sigma < \eta) \wedge (\sigma < \tau \vee \sigma = \tau \vee \tau < \sigma).$$

Σ-reflection axioms: For all Σ-formulæ A:

$$A \to \exists \xi. A^\xi.$$

Σ-induction on the ordinals: For all Σ-formulæ $A(\xi)$,

$$\bigl(\forall \xi. (\forall \eta < \xi. A(\eta)) \to A(\xi)\bigr) \to \forall \xi. A(\xi).$$

Substitution rule: For A an $\mathcal{L}^1$-formula, and $B(x)$ an $\mathcal{L}^1_{\text{On}}$-formula:

$$\frac{A(U)}{A(\{x \mid B(x)\})} \; (\text{SUBST})$$

We need a restriction on A in the substitution rule because the predicates $P_\mathfrak{A}$ will depend on U in general (since the operators $\mathfrak{A}$ may contain U).

Now, $\mathcal{U}(\text{ID}_1)$ embeds into this system, exactly the same way that $\mathcal{U}(\text{NFA})$ embeds into $\text{PA}_\Omega^+ + (\text{Subst})$ (FEFERMAN AND STRAHM 2000). In fact, it is a bit simpler since we are using the PCA version of unfolding, so Δ-induction would suffice.[2]

---

[2] The PCA version of FEFERMAN AND STRAHM (2000) was sketched in FEFERMAN AND STRAHM (2010).

The embedding

The embedding in $(\mathrm{ID}_1)^+_{\mathrm{On}} + (\mathrm{Subst})$ is accomplished by interpreting operations as indices for partial recursive functions with pairing coded in the usual primitive recursive way, and then interpreting Π and $\in$ via an inductive predicate. In fact, to ensure positivity, we need to simultaneously define Π, $\in$ and $\bar{\in}$ (the complement of $\in$). The clauses mirror exactly the defining axioms, so we elide them.

With this inductive encoding, Π, $\in$ and $\bar{\in}$ are Σ-formulæ, whereas the stages Π^α, $\in^\alpha$ and $\bar{\in}^\alpha$ are Δ-formulæ. The complementarity of $\in$ and $\bar{\in}$ follows by Δ-induction on the ordinals as in FEFERMAN AND STRAHM (2000). The predicate axioms are readily seen to hold using just the property of the operator and the complementarity.

Thus we see that with this version of $\mathcal{U}(\mathrm{ID}_1)$, we only need Δ-induction, and not Σ-induction as in the multi-sorted version with least-fixed point recursion operators. However, we shall prove the upper bound for the intermediate system with Σ-induction, partly as a matter of general interest, and partly because this would then also allow a reduction of the LFP version of unfolding.

3.2 INFINITARY LANGUAGES

The language of the previous section, $\mathcal{L}^1_{\mathrm{On}}$, is not suitable for proof-theoretic treatment since the inductive predicate of ID_1, $I_\mathcal{O}$, is not analyzed. And besides, we need to add constants for ordinals. (Since we use operator-controlled derivations, we do not use notations for ordinals, but rather actual ordinals, and let the operators ensure there are enough *gaps* for collapsing. A finitary treatment can then be obtained along the lines of TUPAILO (2000) and BUCHHOLZ (2001).)

The language $\mathcal{L}^1_\infty$ modifies the language $\mathcal{L}^1_{\mathrm{On}}$ by including ordinal constants and by replacing the predicate $I_\mathcal{O}$ with a binary predicate, which we shall also call $I_\mathcal{O}$. This represents the stages of build-up of the original $I_\mathcal{O}$, where the first entry indicates the stage. Just as for the $P_\mathfrak{A}$, we write $I^\alpha_\mathcal{O}(x)$ instead of $I_\mathcal{O}(\alpha, x)$, and we put $I^{<\alpha}_\mathcal{O}(x) := \exists \beta < \alpha . I^\beta_\mathcal{O}(x)$.

Note that the predicates $P_\mathfrak{A}$ are still indexed by operator forms $\mathfrak{A}(X, \vec{x})$ in the language of ID_1. But this language embeds in $\mathcal{L}^1_\infty$ by interpreting $I_\mathcal{O}(x)$ as $I^{<\Omega}_\mathcal{O}(x)$.

The ordinal terms of $\mathcal{L}^1_\infty$ are the ordinal variables and the ordinal constants. The formulæ of $\mathcal{L}^1_\infty$ are generated as follows:
- Every literal is a formula.
- If A and B are formulæ, then $A \wedge B$ and $A \vee B$ are formulæ.

- If $A(x)$ is a formula with a distinguished free number variable x, then $\exists x.\, A(x)$ and $\forall x.\, A(x)$ are formulæ.
- If $A(\xi)$ is a formula with a distinguished ordinal variable ξ, then $\exists \xi.\, A(\xi)$, $\exists \xi < \theta.\, A(\xi)$, $\forall \xi.\, A(\xi)$ and $\forall \xi < \theta.\, A(\xi)$ are formulæ (where θ is an ordinal term).

Note that $\mathcal{L}^1_\infty$ is a Tait-style language, so negation is defined using de Morgan's laws and double negation elimination.

Fragments

We need some fragments of $\mathcal{L}^1_\infty$ to do the proof-theoretic analysis of $(\mathrm{ID}_1)^+_{\mathrm{On}}+(\mathrm{Subst})$. Common to them is a restriction on the appearance of $I_\mathcal{O}$-literals to those needed to interpret the schematic inductive definition axioms. We say that an occurrence of $I_\mathcal{O}$ is *good* if either

- it is $(\neg)I_\mathcal{O}^\alpha(s)$ with $\alpha < \Omega$ an ordinal constant, or
- it occurs as part of $(\neg)I_\mathcal{O}^{<\alpha}(s)$ with $\alpha \leq \Omega$ an ordinal constant.

Let now $\mathcal{L}^{1,\mathrm{r}}_\infty$ consist of the fragment of $\mathcal{L}^1_\infty$-formulæ
- with only good occurrences of $I_\mathcal{O}$, and
- without free variables and without unbounded ordinal quantifiers.

This fragment is suitable for locally predicative cut-elimination.

Let $\mathcal{L}^{1,\mathrm{c}}_\infty$ be the fragment of $\mathcal{L}^1_\infty$ consisting of formulæ
- with only good occurrences of $I_\mathcal{O}$, and
- without free number variables, and
- with all ordinal constants occurring as part of good occurrences of $I_\mathcal{O}$.

This fragment is suitable for embedding the intermediate system. It contains just enough infinitary notions to prove the inductive definition axioms, but no more.

For $\mathcal{L}^{1,\mathrm{c}}_\infty$ we define the notions of Δ-, Σ- and Π-formulæ as for $\mathcal{L}^1_{\mathrm{On}}$.

Finally, let $\mathcal{L}^{1,\mathrm{rc}}_\infty := \mathcal{L}^{1,\mathrm{r}}_\infty \cap \mathcal{L}^{1,\mathrm{c}}_\infty$. Note that this fragment has no ordinal literals $<$ and $=$, as these would occur in scope of a bounded ordinal quantifier, but these only appear in good occurrences of $I_\mathcal{O}$.

Ranks

DEFINITION 3.2.1. *For an $\mathcal{L}_\infty^{1,r}$-formula A define the rank of A, rk(A), as follows:*

$$\begin{aligned}
\text{rk}(A) &:= 0 \quad \text{for } A \text{ a number literal, including } (\neg)U(s) \\
\text{rk}(\alpha < \beta) &:= \text{rk}(\alpha \not< \beta) := \text{rk}(\alpha = \beta) := \text{rk}(\alpha \neq \beta) := \Omega + 1 \\
\text{rk}(I_\mathcal{O}^\alpha(s)) &:= \text{rk}(\neg I_\mathcal{O}^\alpha(s)) := \omega \cdot (\alpha + 1) \\
\text{rk}(P_\mathfrak{A}^\alpha(\vec{s})) &:= \text{rk}(\neg P_\mathfrak{A}^\alpha(\vec{s})) := \Omega + \omega \cdot (\alpha + 1) \\
\text{rk}(A \vee B) &:= \text{rk}(A \wedge B) := \max\{\text{rk}(A), \text{rk}(B)\} + 1 \\
\text{rk}(\exists x.\, G(x)) &:= \text{rk}(\forall x.\, G(x)) := \text{rk}(G(0)) + 1 \\
\text{rk}(\exists \xi < \alpha.\, G(\xi)) &:= \text{rk}(\forall \xi < \alpha.\, G(\xi)) := \sup_{\beta < \alpha} \{\text{rk}(G(\beta)) + 1\}
\end{aligned}$$

Note that rk$(I_\mathcal{O}^{<\Omega}(s)) = \Omega$, and all formulæ with rank Ω are of the form $\exists \xi < \Omega.\, G(\xi)$ or $\forall \xi < \Omega.\, G(\xi)$ for some $G(\xi)$ with rk$(G(0)) < \Omega$.

For an $\mathcal{L}_\infty^1$ formula A, we let par(A) (*parameters of A*) denote the set of ordinal constants occurring in A. If α is an ordinal constant, then we set par$(\alpha) = \{\alpha\}$.

Since $\sup_{\beta < \alpha}\{\beta + 1\} = \alpha$ for $\alpha \in$ Lim, we have by an easy induction that if all elements of par(A) can be represented by terms in our notation system, then so can rk(A), and furthermore we can compute it by a primitive recursive function.

We have the following analogue of STRAHM (2000, Lemma 6):

LEMMA 3.2.2. *For the inductive operator form $\mathcal{O}$ of* ID$_1$, *and any operator form $\mathfrak{A}$ of $\mathcal{L}_{\text{On}}^1$, and any ordinal α we have*
1. rk$(\mathcal{O}(I_\mathcal{O}^{<\alpha}, s)) <$ rk$(I_\mathcal{O}^\alpha)$.
2. rk$(\mathfrak{A}(P_\mathfrak{A}^{<\alpha}, \vec{s})) <$ rk$(P_\mathfrak{A}^\alpha(\vec{s}))$.

For any $\mathcal{L}_\infty^{1,r}$-formula A with par$(A) \subseteq \alpha$, *we have* rk$(A) < \Omega + \omega \cdot (\alpha + 1)$.

Characteristic sequences

We can organize the infinitary calculi using the notion of characteristic sequence, CS, of sub-formulæ for certain formulæ of $\mathcal{L}_\infty^1$ (cf. the analogous notion for ramified set theory in, e.g., POHLERS (1998)). A closed *primitive* literal (one not involving $I_\mathcal{O}$, $P_\mathfrak{A}$, nor U) is either true or false. If it is true, we say it is of $\bigwedge$-type, if false, of $\bigvee$-type. Furthermore, $A \vee B$, $\exists x.\, G(x)$ and $\exists \xi < \alpha.\, G(\alpha)$ are of $\bigvee$-type, while $A \wedge B$, $\forall x.\, G(x)$

and $\forall \xi < \alpha. G(x)$ are of $\bigwedge$-type. Other formulæ are not given a type. Then we put:

$$\begin{aligned}
\text{CS}(A) &:= \emptyset \quad \text{if } A \text{ is a closed primitive literal} \\
\text{CS}(A \vee B) &:= \{A, B\} \\
\text{CS}(\exists x. G(x)) &:= \{G(s) \mid s \text{ a closed number term}\} \\
\text{CS}(\exists \xi < \alpha. G(\xi)) &:= \{G(\beta) \mid \beta < \alpha\}
\end{aligned}$$

For F of $\bigwedge$-type we put $\text{CS}(F) := \{\neg G \mid G \in \text{CS}(\neg F)\}$.

If F is of $\bigvee$-type or $\bigwedge$-type but is not a disjunction or conjunction, then every $G \in \text{CS}(F)$ is of the form $H(t)$ for some (number or ordinal) term t. We define $t_F(G) := t$, considered as an ordinal. For $F = A_0 \circ A_1$ ($\circ = \vee, \wedge$), we define $t_F(A_i) := i$. We note that $\bigvee$-type and $\bigwedge$-type are preserved under substitution, and that $\text{CS}(F)$ as well as t_F commute with substitution.

3.3 THE STRATEGY

The general strategy of the upper bound proof is to embed our system for ID_1 with ordinals in an infinitary system T, based on the language $\mathcal{L}^{1,c}_\infty$, that straightforwardly interprets $(\text{ID}_1)^+_{\text{On}}$, using rules for Σ-reflection and Σ-induction on $\in$. The restricted formulation of the axioms and rules of T enables us to prove a partial cut elimination theorem.

T is then related to a more standard infinitary calculus T^r (r for restricted), based on the restricted language $\mathcal{L}^{1,r}_\infty$ which does not allow free variables nor unbounded quantification. This calculus admits a locally predicative cut elimination theorem and a collapsing theorem for the Σ-fragment.

Due to fact that we are at the level of impredicative proof theory, we will need operator-controlled derivations, and we will need to investigate how the operators behave when shifting between systems.

We shall use judgments of the form $S, \mathcal{H} \vdash^\alpha_\rho \Gamma$, indicating that the sequent Γ is derivable using the system S with a derivation of height $\leq \alpha$ with cut degrees $< \rho$ and controlled by the operator $\mathcal{H}$ (the notion of degree will of course depend on S).

Acceptable operators

Let us recall the standard notions needed for operator controlled derivations (cf. BUCHHOLZ (1992) and POHLERS (2009)). An *operator* is a map $\mathcal{H}: \text{Pow}(\text{On}) \to$

Pow(On). We define:

$$\alpha \in \mathcal{H} :\Leftrightarrow \alpha \in \mathcal{H}(\emptyset),$$
$$M \subseteq \mathcal{H} :\Leftrightarrow M \subseteq \mathcal{H}(\emptyset),$$
$$\mathcal{H} \subseteq M :\Leftrightarrow \mathcal{H}(\emptyset) \subseteq M,$$
$$\mathcal{H} \subseteq \mathcal{H}' :\Leftrightarrow \forall X.\, \mathcal{H}(X) \subseteq \mathcal{H}'(X),$$
$$\mathcal{H}[X] := \lambda Y.\, H(X \cup Y).$$

We say that $\mathcal{H}$ is *Cantorian-closed* if for any $X \subseteq \mathrm{On}$ and any list of ordinals with $\xi_1 \geq \cdots \geq \xi_n$, we have

$$\omega^{\xi_1} + \cdots + \omega^{\xi_n} \in \mathcal{H}(X) \leftrightarrow \{\xi_1, \ldots, \xi_n\} \subseteq \mathcal{H}(X).$$

DEFINITION 3.3.1. An operator $\mathcal{H}$ is *acceptable* if it satisfies the following conditions:
- $0, \Omega \in \mathcal{H}$.
- $\mathcal{H}$ is Cantorian-closed.
- $\forall X.\, X \subseteq \mathcal{H}(X)$.
- $\forall X, Y.\, X \subseteq \mathcal{H}(Y) \to \mathcal{H}(X) \subseteq \mathcal{H}(Y)$ (in particular, $\mathcal{H}$ is idempotent).

The collapsing result for the restricted system that we will treat of in the next section is stated in terms of a particular hierarchy of acceptable operators, $\mathcal{H}_\alpha$, which mirrors the ordinal notation system. Here we recall the definition from BUCHHOLZ (1992):

DEFINITION 3.3.2. For γ any ordinal, let $\mathcal{H}_\gamma$ be the operator that maps $X \subseteq \mathrm{On}$ to $\bigcap \{\mathrm{Cl}(\alpha, \beta) \mid X \subseteq \mathrm{Cl}(\alpha, \beta) \wedge \gamma < \alpha\}$. Note that $\mathcal{H}_\gamma(\emptyset) = \mathrm{Cl}(\gamma + 1, 0) = \mathrm{Cl}_\Omega(\gamma + 1)$.

In particular, all the $\mathcal{H}_\alpha$ are closed under the binary Veblen function, in addition to being acceptable operators.

3.4 THE RESTRICTED INFINITARY CALCULUS

The restricted system T is formulated as a Tait-style calculus for finite sets of $\mathcal{L}^{1,\mathrm{r}}_\infty$-formulæ.

If Θ is a set of $\mathcal{L}^1_\infty$-expressions and $\mathcal{H}$ is an operator, then we set $\mathcal{H}[\Theta] := \mathcal{H}[\mathrm{par}(\Theta)]$.

DEFINITION 3.4.1. The judgment $T^r, \mathcal{H} \mid\frac{\alpha}{\rho} \Delta$ holds if $\text{par}(\Delta) \cup \{\alpha\} \subseteq \mathcal{H}$ and one of the following conditions is satisfied:

(Atom) There are equivalent closed number terms s, t such that $\{\neg U(s), U(t)\} \subseteq \Delta$.

($\bigwedge$) There is a sentence $F \in \Delta$ of $\bigwedge$-type such that $T^r, \mathcal{H}[t_F(G)] \mid\frac{\alpha_G}{\rho} \Delta, G$ and $\alpha_G < \alpha$ for all $G \in \text{CS}(F)$.

($\bigvee$) There is a sentence $F \in \Delta$ of $\bigvee$-type such that $T^r, \mathcal{H} \mid\frac{\alpha_G}{\rho} \Delta, G$ and $t_F(G), \alpha_G < \alpha$ for some $G \in \text{CS}(F)$.

$((\neg)\text{IO}_{\mathcal{O}})$ There is a closed number term s and an ordinal term β with $(\neg)I_{\mathcal{O}}^\beta(s) \in \Delta$, such that $T^r, \mathcal{H} \mid\frac{\alpha_0}{\rho} \Delta, (\neg)\mathcal{O}(I_{\mathcal{O}}^{<\beta}, s)$ and $\alpha_0 < \alpha$.

$(\text{Ref}_{\mathcal{O}})$ There is a closed number term s with $I_{\mathcal{O}}^{<\Omega}(s) \in \Delta$, such that $T^r, \mathcal{H} \mid\frac{\alpha_0}{\rho} \Delta, \mathcal{O}(I_{\mathcal{O}}^{<\Omega}, s)$ and $\alpha_0 < \alpha$.

$((\neg)\text{IO}_{\mathfrak{A}})$ There is a sequence of closed number terms $\vec{s}$ and an ordinal term β with $(\neg)P_{\mathfrak{A}}^\beta(\vec{s}) \in \Delta$, such that $T^r, \mathcal{H} \mid\frac{\alpha_0}{\rho} \Delta, (\neg)\mathfrak{A}(I_{\mathfrak{A}}^{<\beta}, \vec{s})$ and $\alpha_0 < \alpha$.

(Cut) There is a formula A with $\text{rk}(A) < \rho$, $T^r, \mathcal{H} \mid\frac{\alpha_0}{\rho} \Delta, A$ and $T^r, \mathcal{H} \mid\frac{\alpha_0}{\rho} \Delta, \neg A$ for some $\alpha_0 < \alpha$.

Note that the inductive operator rules $((\neg)\text{IO}_{\mathcal{O}})$ and $((\neg)\text{IO}_{\mathfrak{A}})$ come in pairs, one with $\neg$ and the other without.

The following results are proved very much along the same lines as those of POHLERS (1998). However, since our setup is different, we will comment on the places where things are different.

LEMMA 3.4.2 (Inversion Lemma). *If $T^r, \mathcal{H} \mid\frac{\alpha}{\rho} \Delta, F$ where F is a sentence of $\bigwedge$-type, then $T^r, \mathcal{H}[t_F(G)] \mid\frac{\alpha}{\rho} \Delta, G$ for all $G \in \text{CS}(F)$.*

LEMMA 3.4.3 (Reduction Lemma). *If $\text{rk}(F) = \rho \neq \Omega$, and we have both $T^r, \mathcal{H} \mid\frac{\alpha}{\rho} \Gamma, \neg F$ as well as $T^r, \mathcal{H} \mid\frac{\beta}{\rho} \Delta, F$, then $T^r, \mathcal{H} \mid\frac{\alpha \# \beta}{\rho} \Gamma, \Delta$.*

Proof. The proof is by induction on $\alpha \# \beta$. If either F or $\neg F$ is not a main formula of the last inference of the respective derivation, then we are done by induction hypothesis. So assume that both are main formulæ.

First suppose that $F \equiv U(s)$. Since F and $\neg F$ are main formulæ, the last inference in both derivations is (ATOM), so we have $U(t) \in \Gamma$ and $\neg U(r) \in \Delta$ with r, s, t numerically equivalent closed number terms. But then Γ, Δ can be concluded by (ATOM).

Next suppose that the last inferences have inductive operators as main formulæ

25

(either $I_\mathcal{O}$ or $P_\mathfrak{A}$). (REF$_\mathcal{O}$) is ruled out by supposition, so we can just cut the hypothesis sequents which have matching formulæ of lower rank.

Otherwise, F is assigned a type (since we're in the restricted system with no free individual variables), and we can assume without loss of generality that $F \in \bigvee$-type. Then we can show that $\alpha + \beta$ is a bound for the reduced derivation, following POHLERS (1998, Lemma 3.4.3.5). $\square$

LEMMA 3.4.4 (Predicative Elimination Lemma). *Let $\mathcal{H}$ be an operator that is closed under $\alpha, \beta \mapsto \varphi_\alpha \beta$, and suppose* $T^r, \mathcal{H} \mid\!\frac{\alpha}{\beta + \omega^\rho}\ \Delta$, *where $\Omega \notin [\beta, \beta + \omega^\rho)$ and $\rho \in \mathcal{H}$. Then*
$$T^r, \mathcal{H} \mid\!\frac{\varphi_\rho \alpha}{\beta}\ \Delta.$$

For the next results we need the notion of a Σ^ξ-sentence of $\mathcal{L}_\infty^{1,r}$. Such is obtained from a closed Σ-formula A of $\mathcal{L}_\infty^1$ with par$(A) \subseteq \xi$ by bounding the existential quantifiers by ξ. Prominently, $I_\mathcal{O}^{<\Omega}(s)$ and $\mathcal{O}(I_\mathcal{O}^{<\Omega}, s)$ are both Σ^Ω-sentences.

If A^ξ is a Σ^ξ-sentence and $\xi \leq \eta$, then A^η is the Σ^η-sentence we obtain by taking η as the bounds instead of ξ.

LEMMA 3.4.5 (Lifting Lemma). *Suppose A^ξ is a Σ^ξ-sentence and $\xi \leq \eta$. Then we have:*
1. $T^r, \mathcal{H} \mid\!\frac{\alpha}{\rho}\ \Delta, A^\xi$ *implies* $T^r, \mathcal{H}[\eta] \mid\!\frac{\alpha}{\rho}\ \Delta, A^\eta$.
2. $T^r, \mathcal{H} \mid\!\frac{\alpha}{\rho}\ \Delta, I_\mathcal{O}^\xi(s)$ *implies* $T^r, \mathcal{H}[\eta] \mid\!\frac{\alpha}{\rho}\ \Delta, I_\mathcal{O}^\eta(s)$.
3. $T^r, \mathcal{H} \mid\!\frac{\alpha}{\rho}\ \Delta, P_\mathfrak{A}^\xi(s)$ *implies* $T^r, \mathcal{H}[\eta] \mid\!\frac{\alpha}{\rho}\ \Delta, P_\mathfrak{A}^\eta(s)$.

LEMMA 3.4.6. *If* $T^r, \mathcal{H} \mid\!\frac{\alpha}{\rho}\ \Delta, \forall \xi < \Omega. G(\xi)$ *and* $\beta \in \mathcal{H} \cap \Omega$, *then* $T^r, \mathcal{H} \mid\!\frac{\alpha}{\rho}\ \Delta, \forall \xi < \beta. G(\xi)$.

The two preceding Lemmata are both proved by straightforward induction on α.

THEOREM 3.4.7 (Boundedness Theorem). *If* $T^r, \mathcal{H} \mid\!\frac{\alpha}{\rho}\ \Delta, F^\Omega$ *for a Σ^α-sentence F^α with $\alpha < \Omega$, then* $T^r, \mathcal{H} \mid\!\frac{\alpha}{\rho}\ \Delta, F^\alpha$.

Proof sketch. The proof is by induction on α; cf. POHLERS (1998, Theorem 3.4.3.7). The only different case is when $F^\Omega \equiv I_\mathcal{O}^{<\Omega}(s)$ in the case of an inference (REF$_\mathcal{O}$) with premise
$$T^r, \mathcal{H} \mid\!\frac{\alpha_0}{\rho}\ \Delta, I_\mathcal{O}^{<\Omega}(s), \mathcal{O}(I_\mathcal{O}^{<\Omega}, s),$$

for some $\alpha_0 < \alpha$. Using positivity of $\mathcal{O}$ we see that $\mathcal{O}(I_\mathcal{O}^{<\Omega}, s)$ is again a Σ^Ω-sentence, so by two uses of the induction hypothesis, we get

$$T^r, \mathcal{H} \vdash^{\alpha_0}_\rho \Delta, I_\Theta^{<\alpha_0}(s), \mathcal{O}(I_\Theta^{<\alpha_0}, s).$$

Now we are done by the Lifting Lemma and an inference (IO$_\Theta$). □

THEOREM 3.4.8 (Collapsing Theorem for T^r). *Let Δ be a set of Σ^Ω-sentences of $\mathcal{L}^{1,r}_\infty$, and assume $T^r, \mathcal{H}_0 \vdash^{\alpha}_{\Omega+1} \Delta$. Then*

$$T^r, \mathcal{H}_{\omega^{\Omega+1+\alpha}} \vdash^{\psi(\omega^{\Omega+1+\alpha})}_{\psi(\omega^{\Omega+1+\alpha})} \Delta.$$

Proof sketch. As in BUCHHOLZ (1992) and later analyses using operator-controlled derivations, we need to strengthen the statement in order to have the induction go through.

For Θ a set of ordinals and γ an ordinal, let $\mathcal{A}(\Theta; \gamma)$ denote the conjunction

$$\text{par}(\Theta) \subseteq \text{Cl}_\Omega(\gamma + 1) \wedge \gamma \in \mathcal{H}_\gamma[\Theta].$$

Then we prove: For all Θ and γ with $\mathcal{A}(\Theta; \gamma)$, if Δ is a set of Σ^Ω-sentences such that $T^r, \mathcal{H}_\gamma[\Theta] \vdash^{\alpha}_{\Omega+1} \Delta$, then $T^r, \mathcal{H}_{\hat{\alpha}}[\Theta] \vdash^{\psi(\hat{\alpha})}_{\psi(\hat{\alpha})} \Delta$ where $\hat{\alpha} := \gamma + \omega^{\Omega+1+\alpha}$.[3]

The proof is by induction on α. Most of the cases are standard, except (ATOM) (which is trivial), (REF$_\Theta$), and (CUT). For (REF$_\Theta$), assume we have $I_\Theta^{<\Omega}(s) \in \Delta$ and an ordinal $\alpha_0 < \alpha$ such that

$$T^r, \mathcal{H}_\gamma[\Theta] \vdash^{\alpha_0}_{\Omega+1} \Delta, \mathcal{O}(I_\Theta^{<\Omega}, s).$$

By the induction hypothesis we have, with $\hat{\alpha}_0 := \gamma + \omega^{\Omega+1+\alpha_0}$,

$$T^r, \mathcal{H}_{\hat{\alpha}_0}[\Theta] \vdash^{\psi(\hat{\alpha}_0)}_{\psi(\hat{\alpha}_0)} \Delta, \mathcal{O}(I_\Theta^{<\Omega}, s).$$

By the Boundedness Theorem, we get

$$T^r, \mathcal{H}_{\hat{\alpha}_0}[\Theta] \vdash^{\psi(\hat{\alpha}_0)}_{\psi(\hat{\alpha}_0)} \Delta, \mathcal{O}(I_\Theta^{<\psi(\hat{\alpha}_0)}, s).$$

---
[3] The set Θ is needed to take care of ($\bigwedge$)-inferences.

Now use inferences (IO$_\Theta$) and ($\bigvee$) to conclude.

In the case of (CUT), we have a sentence A with $\mathrm{rk}(A) \leq \Omega$ and an ordinal α_0 such that
$$T^r, \mathcal{H}_\gamma[\Theta] \frac{\alpha_0}{\Omega+1} \Delta, A \quad \text{and} \quad T^r, \mathcal{H}_\gamma[\Theta] \frac{\alpha_0}{\Omega+1} \Delta, \neg A.$$
If $\mathrm{rk}(A) < \Omega$, then $\mathrm{rk}(A) \in \mathcal{H}_\gamma[\Theta] \cap \Omega \subseteq \psi(\gamma+1) \leq \psi(\gamma + \omega^{\Omega+1+\alpha})$, so we're done by induction hypothesis and (CUT).

If $\mathrm{rk}(A) = \Omega$, we may assume $A \equiv \exists \xi < \Omega. G(\xi)$. To the first premise, we may apply the induction hypothesis. Since $\mathrm{par}(G(0)) \subseteq \mathcal{H}_\gamma[\Theta] \cap \Omega \subseteq \psi(\gamma+1) \leq \psi(\hat{\alpha}_0)$ (with $\hat{\alpha}_0$ as above), we see that $\exists \xi < \psi(\hat{\alpha}_0). G(\xi)$ is a $\Sigma^{\psi(\hat{\alpha}_0)}$-sentence, so we may apply boundedness to get
$$T^r, \mathcal{H}_{\hat{\alpha}_0}[\Theta] \frac{\psi(\hat{\alpha}_0)}{\psi(\hat{\alpha}_0)} \Delta, \exists \xi < \psi(\hat{\alpha}_0). G(\xi).$$

To the second premise, we may apply Lemma 3.4.6 to get
$$T^r, \mathcal{H}_{\hat{\alpha}_0}[\Theta] \frac{\alpha_0}{\Omega+1} \Delta, \forall \xi < \psi(\hat{\alpha}_0). \neg G(\xi).$$

We easily check $\mathcal{A}(\Theta; \hat{\alpha}_0)$ so by induction hypothesis we get
$$T^r, \mathcal{H}_{\gamma'}[\Theta] \frac{\psi(\gamma')}{\psi(\gamma')} \Delta, \forall \xi < \psi(\hat{\alpha}_0). \neg G(\xi),$$
where $\gamma' := \hat{\alpha}_0 + \omega^{\Omega+1+\alpha_0} < \hat{\alpha}$. By (CUT) we obtain
$$T^r, \mathcal{H}_{\gamma'}[\Theta] \frac{\psi(\gamma')+1}{\psi(\gamma')} \Delta,$$
which readily entails
$$T^r, \mathcal{H}_{\hat{\alpha}}[\Theta] \frac{\psi(\hat{\alpha})}{\psi(\hat{\alpha})} \Delta,$$
as desired. $\square$

Completely parallel to the development of POHLERS (2009, Chapter 9), we may prove a Controlled Tautology and Monotonicity result for T^r:

LEMMA 3.4.9 (Controlled Tautology for T^r). *For any acceptable operator $\mathcal{H}$ and numerically equivalent $\mathcal{L}^{1,r}_\infty$-formulæ A and B, if $\mathrm{par}(\Delta, A) \subseteq \mathcal{H}$, then*
$$T^r, \mathcal{H} \frac{2 \cdot \mathrm{rk}(A)}{0} \Delta, \neg A, B.$$

LEMMA 3.4.10 (Monotonicity for T^r). *Let $\mathcal{H}$ be an acceptable operator, and suppose* T^r, $\mathcal{H} \vdash^{\alpha}_{\rho} \Delta, \neg A(s), B(s)$ *for all closed number terms s. If $\mathcal{O}(X, x)$ is an X-positive arithmetical operator form of $\mathcal{L}^0$, then*

$$\mathrm{T}^r, \mathcal{H} \vdash^{\alpha + 2 \cdot \mathrm{rk}(\mathcal{O})}_{\rho} \Delta, \neg \mathcal{O}(A, s), \mathcal{O}(B, s)$$

for all closed number terms s.

LEMMA 3.4.11 (Generalized induction in T^r). *For any acceptable operator $\mathcal{H}$ we have*

$$\mathrm{T}^r, \mathcal{H}[\alpha] \vdash^{\omega \cdot (\alpha+1)}_{0} \neg(\forall x. \mathcal{O}(U, x) \to U(x)), \neg I^{\alpha}_{\mathcal{O}}(s), U(s).$$

This is derived from the previous Lemmata, and from this we in turn obtain by a few logical inferences that

$$\mathrm{T}^r, \mathcal{H} \vdash^{\Omega+3}_{0} (\forall x. \mathcal{O}(U, x) \to U(x)) \to \forall x. I^{<\Omega}_{\mathcal{O}}(x) \to U(x).$$

Thus we see that both of the schematic induction axioms are controlled derivable in T^r (the first one following directly from (REF$_\mathcal{O}$)). Note in particular that we only need good occurrences of $I_\mathcal{O}$.

3.5 THE DERIVED INFINITARY CALCULUS

Like STRAHM (2000) we shall use a derived infinitary calculus T to model (ID$_1$)$^+_{\mathrm{On}}$. Iterating asymmetric interpretation into T^r with substitution back into T will yield the desired upper bound. We have to be very careful with the fragments involved; see Fig. 3.1 on the next page for a diagram showing these steps.

The system T is derived by extending the rules of T^r, but formulated in the language fragment $\mathcal{L}^{1,c}_\infty$. This is precisely what is needed to embed (ID$_1$)$^+_{\mathrm{On}}$ + (Subst) and unravel the inductively defined predicate $I_\mathcal{O}$.

REMARK 3.5.1. It is potentially a source of confusion that we will use two different notions of rank or degree to control cuts in the two systems. A mnemonic device that might help is that the *restricted* system uses only *rank* defined in Section 3.2, while the *derived* system uses *degree* over $\Sigma \cup \Pi$-formulæ.

In the following definition k bounds the *degree* (over the $\Sigma \cup \Pi$-fragment) of cut-formulæ in a derivation.

Definition 3.5.2. The *degree* of a formula F of $\mathcal{L}_\infty^{1,c}$ over the $\Sigma \cup \Pi$-fragment is 0 if $F \in \Sigma \cup \Pi$; otherwise it is $\max\{\mathrm{dg}(A), \mathrm{dg}(B)\} + 1$ if $F \equiv A \circ B$ ($\circ = \vee, \wedge$); it is $\mathrm{dg}(A) + 1$ if $F \equiv Q\xi. A(\xi)$; it is $\mathrm{dg}(A) + 2$ if $F \equiv Q\xi < \alpha. A(\xi)$ ($Q = \exists, \forall$).

Definition 3.5.3. The judgment $T, \mathcal{H} \vdash^{\alpha}_{k} \Delta$ holds if $\mathrm{par}(\Delta) \cup \{\alpha\} \subseteq \mathcal{H}$ and one of the following conditions is satisfied:

(Atom), ($\bigwedge$), ($\bigvee$), ($IO_\mathcal{O}$), ($Ref_\mathcal{O}$), ($IO_\mathfrak{A}$) As in Definition 3.4.1 with T^r replaced by T and ρ replaced by k (and now variables are allowed in the stage rules).

(Ax) There are numerically equivalent Σ-formulæ A and B and either $\{\neg A, B\} \subseteq \Delta$ or $\{\sigma \neq \tau, \neg A(\sigma), B(\tau)\} \subseteq \Delta$, or there is a linearity axiom of $(ID_1)^+_{On}$ in Δ.

(Cut) There is a formula A with $\mathrm{dg}(A) < k$, $T, \mathcal{H} \vdash^{\alpha_0}_{k} \Delta, A$ and $T, \mathcal{H} \vdash^{\alpha_0}_{k} \Delta, \neg A$ for some $\alpha_0 < \alpha$.

($\forall$) There is a formula $A(\xi)$ with ξ fresh such that $(\forall \xi. A(\xi)) \in \Delta$ and $T, \mathcal{H} \vdash^{\alpha_0}_{k} \Delta, A(\xi)$ for some $\alpha_0 < \alpha$.

($\exists$) There is an ordinal variable β and a formula $(\exists \xi. A(\xi)) \in \Delta$ with $T, \mathcal{H} \vdash^{\alpha_0}_{k} \Delta, A(\beta)$ for some $\alpha_0 < \alpha$.

(Σ-Ref) There is a Σ-formula A with $(\exists \xi. A^\xi) \in \Delta$ such that $T, \mathcal{H} \vdash^{\alpha_0}_{k} \Delta, A$ for some $\alpha_0 < \alpha$.

(Σ-Ind) There is a Σ-formula $A(\xi)$ with ξ fresh and $A(\beta) \in \Delta$ for a variable β such that $T, \mathcal{H} \vdash^{\alpha_0}_{k} \Delta, \neg(\forall \eta < \xi. A(\eta)), A(\xi)$ for some $\alpha_0 < \alpha$.

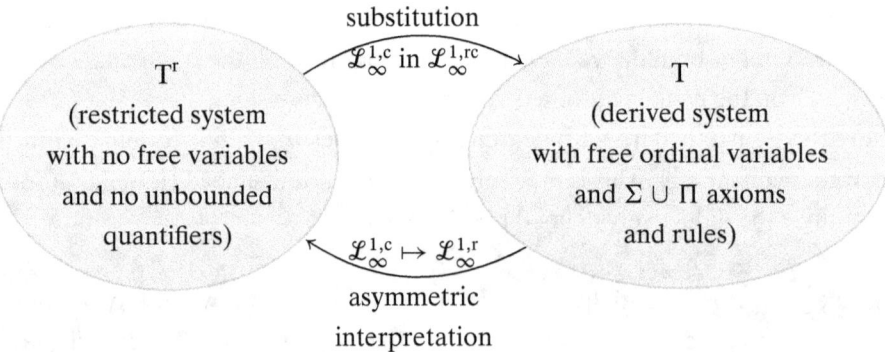

Figure 3.1: Relation between infinitary systems for the upper bound

THEOREM 3.5.4 (Partial cut elimination for T). *For all finite sets Γ of $\mathcal{L}^{1,c}_\infty$-formulæ, and all operators $\mathcal{H}$ closed under $k, \alpha \mapsto 2_k(\alpha)$:*

$$\text{T}, \mathcal{H} \mid\!\frac{\alpha}{k}\, \Gamma \quad \rightarrow \quad \text{T}, \mathcal{H} \mid\!\frac{2_k(\alpha)}{1}\, \Gamma.$$

Proof. This is clear because the main formulæ of all non-logical axioms and rules are in $\Sigma \cup \Pi$. □

Completely analogously to the Tautology Lemma for T^r we have:

LEMMA 3.5.5 (Tautology lemma for T). *We have for all finite sets Γ of $\mathcal{L}^{1,c}_\infty$-formulæ, and all numerically equivalent $\mathcal{L}^{1,c}_\infty$-formulæ A and B, and all acceptable operators $\mathcal{H}$ with $\mathrm{par}(\Gamma, A, B) \subseteq \mathcal{H}$:*

$$\text{T}, \mathcal{H} \mid\!\frac{2 \cdot \mathrm{dg}(A)}{0}\, \Gamma, \neg A, B.$$

3.6 CONNECTING THE SYSTEMS

LEMMA 3.6.1 (Substitution lemma for T^r into T). *Let $\Gamma(U)$ be a finite set of $\mathcal{L}^{1,rc}_\infty$-formulæ with no occurrences of $P_\mathfrak{A}$-literals, and let $B(x)$ be any formula of $\mathcal{L}^{1,c}_\infty$. Assume $\text{T}^r, \mathcal{H} \mid\!\frac{\alpha}{\Omega+1}\, \Gamma(U)$ for some infinite ordinal α. Then we have $\text{T}, \mathcal{H}[\mathrm{par}\, B] \mid\!\frac{\alpha}{<\omega}\, \Gamma(\{x \mid B(x)\})$.*

Proof. The case of (ATOM) is handled by the Tautology Lemma. Because of the restriction on $\Gamma(U)$ and the cut ranks, the entire derivation is free of $P_\mathfrak{A}$-literals (note that the corresponding rules are not invariant under substitution), and in fact stays within the $\mathcal{L}^{1,rc}_\infty$-fragment since any ordinal literals $<$ and $=$ push the rank of a formula above Ω.

Since substitution commutes with characteristic sequences, the rules $(\bigvee)$ and $(\bigwedge)$ are preserved. Since the operator $\mathcal{O}$ is U-free, the rules $((\neg)\text{IO}_\mathcal{O})$ and $(\text{REF}_\mathcal{O})$ are preserved as well. □

Next we will give an asymmetric interpretation of the $\Sigma \cup \Pi$-fragment of T into T^r. Let Γ be a finite set of $\mathcal{L}^{1,c}_\infty$-formulæ, and Λ a finite set of $\mathcal{L}^{1,r}_\infty$-formulæ. We call Λ a (β, γ)-instance of Γ, when Λ arises from Γ by
  (i) replacing each free ordinal variable by an ordinal term less than β.
  (ii) bounding each unbounded universal quantifier by β,
  (iii) bounding each unbounded existential quantifier by γ.

A simple but crucial fact about these instances is persistency for controlled T^r-derivability, which is proved easily like the Lifting Lemma and Lemma 3.4.6.

LEMMA 3.6.2 (Persistency). *Suppose Γ is a finite set of $\mathcal{L}_\infty^{1,c}$-formulæ and Δ is a finite set of $\mathcal{L}_\infty^{1,r}$-formulæ. Let Λ_i (for $i = 0, 1$) be (β_i, γ_i)-instances of Γ with the same terms substituted for free variables, where $\beta_1 < \beta_0$ and $\gamma_1 > \gamma_0$. If*

$$\operatorname{par}(\Delta), \operatorname{par}(\Lambda_0), \operatorname{par}(\Lambda_1) \subseteq \mathcal{H},$$

then

$$T^r, \mathcal{H} \,\Big|\!\frac{\alpha}{\rho}\, \Delta, \Lambda_0 \quad \text{implies} \quad T^r, \mathcal{H} \,\Big|\!\frac{\alpha}{\rho}\, \Delta, \Lambda_1.$$

THEOREM 3.6.3 (Asymmetric interpretation of T into T^r). *Assume Γ is a finite set of $\Sigma \cup \Pi$-formulæ of T so that $T, \mathcal{H} \,\Big|\!\frac{\alpha}{1}\, \Gamma$. Let $\beta \geq \Omega$ be a limit ordinal and put $\gamma := \varphi\alpha(\beta + \beta)$. Then for every (β, γ)-instance Λ of Γ with $\operatorname{par}(\Lambda) \cup \{\beta\} \subseteq \mathcal{H}$ we have $T^r, \mathcal{H} \,\Big|\!\frac{\gamma}{\gamma}\, \Lambda$.*

Proof. Induction on α. We will deal with each case separately.

(ATOM) This is clear.

($\bigwedge$), ($\bigvee$) Note that a formula $F \in \Gamma$ has a corresponding instance $\tilde{F} \in \Lambda$, and for any $G \in \operatorname{CS}(F)$ there is a uniquely corresponding instance $\tilde{G} \in \operatorname{CS}(\tilde{F})$ with $t_{\tilde{F}}(\tilde{G}) = t_F(G)$. Thus the induction hypothesis can be used directly in each case.

(($\neg$)IO$_\mathcal{O}$), (REF$_\mathcal{O}$) Since all occurrences of $I_\mathcal{O}$ are good, these rules are unchanged from Γ to Λ, so the induction hypothesis applies immediately.

(($\neg$)IO$_\mathfrak{A}$) The only possible free ordinal variable is the stage variable that becomes instantiated. Since $\operatorname{par}(\Lambda) \subseteq \mathcal{H}$, we can apply the induction hypothesis.

(Ax) Suppose $\{\neg A, B\} \subseteq \Gamma$ with A and B numerically equivalent Σ-formulæ of $\mathcal{L}_\infty^{1,c}$. Letting $\tilde{A}^\beta$ and $\tilde{B}^\gamma$ be the corresponding instances in Λ, we see that $\operatorname{par}(\tilde{A}^\beta) \subseteq \beta \cup \{\Omega\} \subseteq \beta + 1$ by definition of the language fragments (in particular because of the restriction to good occurrences of $I_\mathcal{O}$). By Lemma 3.2.2 we then get $\operatorname{rk}(\tilde{A}^\beta) < \Omega + \omega \cdot (\beta + 2) < \omega^{\beta + \beta} \leq \varphi\alpha(\beta + \beta)$. Now we are done using the Tautology Lemma and Persistency.

The case $\{\sigma \neq \tau, \neg A(\sigma), B(\tau)\}$ is similar, and the linearity axioms become true closed primitive literals.

(CUT) We handle cut exactly as in other asymmetric interpretations, taking cases according to the shape of the cut formula. If the main connective is not an unbounded quantifier we can use the induction hypotheses, because $\beta \geq \Omega$

implies that the rank is less than γ. In the case of an unbounded quantifier we have

$$T, \mathcal{H} \models^{\alpha_0}_{1} \Gamma, \forall \xi. \neg A(\xi) \quad \text{and} \quad T, \mathcal{H} \models^{\alpha_0}_{1} \Gamma, \exists \xi. A(\xi),$$

where A is a Δ-formula. The given $(\beta, \varphi\alpha(\beta + \beta))$-instance Λ of Γ determines instantiations for some, though perhaps not all free variables of A. Instantiate the remaining by 0 to get a compatible instantiation $\tilde{A}$. Put $\gamma_0 := \varphi\alpha_0(\beta + \beta)$. By induction hypothesis we get (for the (β, γ_0)-instance Λ_0)

(3.1) $$T^r, \mathcal{H} \models^{\gamma_0}_{\gamma_0} \Lambda_0, \exists \xi < \gamma_0. \tilde{A}(\xi).$$

By induction hypothesis for the other judgment, using γ_0 as the universal bound, we get for the (γ_0, γ_1)-instance Λ_1 (where $\gamma_1 := \varphi\alpha_0\gamma_0$)

(3.2) $$T^r, \mathcal{H} \models^{\gamma_1}_{\gamma_1} \Lambda_1, \forall \xi < \gamma_0. \neg \tilde{A}(\xi).$$

Persistency applied to (3.1) and (3.2) now yields

$$T^r, \mathcal{H} \models^{\gamma_0}_{\gamma_0} \Lambda, \exists \xi < \gamma_0. \tilde{A}(\xi) \quad \text{and} \quad T^r, \mathcal{H} \models^{\gamma_1}_{\gamma_1} \Lambda, \forall \xi < \gamma_0. \neg \tilde{A}(\xi).$$

Now we're done using a (Cut)-inference in T^r, as long as we can verify that $\operatorname{rk}(\exists \xi < \gamma_0. \tilde{A}(\xi)) < \varphi\alpha(\beta + \beta)$. But this follows from Lemma 3.2.2.

The remaining cases are handled exactly as by Strahm (2000, Theorem 13). □

3.7 REDUCTION

Closed formulæ of the languages $\mathcal{L}^1$ and $\mathcal{L}^1_{\mathrm{On}}$ embed into $\mathcal{L}^{1,r}_\infty$ and $\mathcal{L}^{1,c}_\infty$ as described in Section 3.2. Here, the translations shall be denoted by a $\star$.

The goal of this section is to analyze the derivations in $(\mathrm{ID}_1)^+_{\mathrm{On}} + (\mathrm{Subst})$ in terms of the infinitary calculi of the preceding sections to establish the upper bound. As in Strahm (2000), we analyze the fragments $(\mathrm{ID}_1)^+_{\mathrm{On}} + (\mathrm{Subst})^{\leq n}$ where the substitution rule is used at most n times.

A fundamental sequence for $\psi(\Gamma_{\Omega+1})$ is obtained as follows: let $\xi_0 := \varepsilon_{\Omega+1}$, and $\xi_{n+1} := \varphi\xi_n 0$.[4] Then $\sup_{n<\omega} \psi(\xi_n) = \psi(\Gamma_{\Omega+1})$.

---
[4] We start at $\varepsilon_{\Omega+1}$ instead of $\Omega+1$ in order to get a smoother formulation of the reduction theorem.

THEOREM 3.7.1 (Reduction of $(\mathrm{ID}_1)^+_{\mathrm{On}}$ + (Subst)). *Let C be a formula of $\mathcal{L}^1_{\mathrm{On}}$, and let A be a closed formula of $\mathcal{L}^1$. Then we have for all natural numbers n, and all acceptable operators $\mathcal{H}$:*

1. $(\mathrm{ID}_1)^+_{\mathrm{On}}$ + (Subst)$^{\leq n} \vdash C \quad \rightarrow \quad \mathrm{T}, \mathcal{H} \vdash^{<\xi_{2n}}_{1} C^\star$.

2. $(\mathrm{ID}_1)^+_{\mathrm{On}}$ + (Subst)$^{\leq n} \vdash A \quad \rightarrow \quad \mathrm{T}^\mathrm{r}, \mathcal{H} \vdash^{<\xi_{2n+2}}_{\Omega+1} A^\star$.

Proof. We follow STRAHM (2000) and prove 1. and 2. together by induction on n.

Base case ($n = 0$): As discussed above, the defining axioms for $(\mathrm{ID}_1)^+_{\mathrm{On}}$ have proofs in T of height $\Omega + n$ for some $n < \omega$. Applying partial cut elimination for T then yields 1. Asymmetric Interpretation followed by Predicative Cut Elimination for T^r now gives 2.

Step case (inferring statement for $n + 1$): This is done by induction on derivations with every case except (SUBST) being trivial. So assume $C \equiv A(B)$ is proved by (SUBST) applied to the premise $A(U)$. By induction hypothesis, $\mathrm{T}^\mathrm{r}, \mathcal{H} \vdash^{<\xi_{2n+2}}_{\Omega+1} A^\star(U)$. Note that $A^\star$ is free of $P_{\mathfrak{A}}$-literals and $\mathrm{par}(B^\star) \subseteq \{\Omega\} \subseteq \mathcal{H}$, so by the Substitution Lemma for T^r into T, we have $\mathrm{T}, \mathcal{H} \vdash^{<\xi_{2n+2}}_{<\omega} A^\star(B^\star)$. By Partial Cut Elimination for T, $\mathrm{T}, \mathcal{H} \vdash^{<\xi_{2n+2}}_{1} A^\star(B^\star)$ since ξ_{2n+2} is closed under $\xi \mapsto 2^\xi$.

If C is in $\mathcal{L}^1$, we can again use Asymmetric Interpretation followed by Predicate Cut Elimination for T^r to get 2. □

COROLLARY 3.7.2. $|(\mathrm{ID}_1)^+_{\mathrm{On}} + (\mathrm{Subst})| \leq \psi(\Gamma_{\Omega+1})$.

Proof. Let A be the $\mathcal{L}^0$-formula $\mathrm{TI}(U, \prec)$ for a primitive recursive binary relation $\prec$. If $(\mathrm{ID}_1)^+_{\mathrm{On}}$ + (Subst) $\vdash A$, then only finitely many instances of (Subst) were used, so by the previous theorem, we get $\mathrm{T}^\mathrm{r}, \mathcal{H}_0 \vdash^{<\xi_n}_{\Omega+1} A^\star$ for some n. By the Collapsing Theorem for T^r, we then get

$$\mathrm{T}^\mathrm{r}, \mathcal{H}_{\xi_{n+1}} \vdash^{<\psi(\xi_{n+1})}_{<\psi(\xi_{n+1})} A^\star.$$

Since $\psi(\Gamma_{\Omega+1})$ is strongly critical, we can now use the Predicative Cut Elimination Theorem for T^r repeatedly to conclude $\mathrm{T}^\mathrm{r}, \mathcal{H}_{\xi_{n+1}} \vdash^{<\psi(\Gamma_{\Omega+1})}_{0} A^\star$. Now we can forget about the operator and see that $\vdash^{<\psi(\Gamma_{\Omega+1})}_{0} A^\star$ in a standard infinitary system for number theory. This implies that the order-type of $\prec$ is bounded by $\psi(\Gamma_{\Omega+1})$; cf. POHLERS (2009, Chapter 6)). □

Chapter 4

LOWER BOUNDS

4.1 PRELIMINARIES

In this section we recall the notion of *accessible part* of a binary relation $\prec$. This is the inductively defined subset of the field of $\prec$ determined by the operator

$$\lambda X. \lambda x. \forall y \prec x. y \in X.$$

The least fixed point is denoted $\mathrm{Acc}_\prec$, and $|s|_{\mathrm{Acc}_\prec}$ denotes the inductive norm of s with respect to $\mathrm{Acc}_\prec$.

We define:

$$\mathrm{Prog}_\prec(X) :\Leftrightarrow \forall x. (\forall y \prec x. y \in X) \to x \in X,$$
$$\mathrm{TI}_\prec(s, X) :\Leftrightarrow \mathrm{Prog}_\prec(X) \to \forall x \prec s. x \in X.$$

LEMMA 4.1.1. *If $\ll$ is a subrelation of $\prec$ (i.e., for all x, y we have that $x \ll y$ implies $x \prec y$), then $\mathrm{Prog}_\ll(X)$ implies $\mathrm{Prog}_\prec(X)$.*

COROLLARY 4.1.2. *If $\ll$ is a subrelation of $\prec$, then $\mathrm{TI}_\prec(s, X)$ implies $\mathrm{TI}_\ll(s, X)$.*

4.2 OUTLINE

We will prove that there is a primitive recursive relation $<_0$ such that for every $\alpha < \psi(\Gamma_{\Omega+1})$ we get $\mathcal{U}(\mathrm{ID}_1) \vdash n \in \mathrm{Acc}_{<_0}$ and $|n|_{\mathrm{Acc}_{<_0}} = \alpha$ for some n.

We will do this by combining the condensing argument of POHLERS (2009, Section 9.6.2) with the strategy of FEFERMAN AND STRAHM (2000, Section 5.1).

In the rest of this chapter we will talk exclusively about ordinal notations in OT, but refer to them as ordinals for short.

DEFINITION 4.2.1. For ordinals α, β define

$$\alpha <_0 \beta :\Leftrightarrow \alpha < \beta < \Omega.$$

Let $\mathcal{A}cc := \text{Acc}_{<_0} \cap \Omega$ and $M := \{\alpha \mid SC(\alpha) \cap \Omega \subseteq \mathcal{A}cc\}$ and define

$$\alpha <_1 \beta :\Leftrightarrow \alpha < \beta \wedge \alpha \in M \wedge \beta \in M.$$

Note that the $<_0$ relation is arithmetical, whereas the $<_1$ relation refers to the least fixed point, $\text{Acc}_{<_0}$. Instead of defining $\text{Acc}_{<_0}$ as a fixed point, we could have directly defined $\mathcal{A}cc$ as the least fixed point of the operator

$$\lambda X. \lambda \xi. \xi < \Omega \wedge \forall \eta < \xi. \eta \in X.$$

Then $\text{Acc}_{<_0} = \mathcal{A}cc \cup \{\xi \mid \xi \not< \Omega\}$. So $\text{Acc}_{<_0}$ and $\mathcal{A}cc$ are interdefinable.

Also note that in the standard model, M is the set of all ordinal notations, so there $<_1$ coincides with $<$. But since we can only prove that certain initial segments are well-founded, the $<_1$ relation is what allows the lower bound proof to go through.

In Section 4.4, we shall prove transfinite induction in the sense of $<_1$ for every initial segment of $\Gamma_{\Omega+1}$. Then in Section 4.5, we use a standard condensation argument to prove transfinite induction in the sense of $<_0$ for every initial segment of $\psi(\Gamma_{\Omega+1})$.

4.3 THE ACCESSIBLE PART

From now on we write $\text{TI}_i(s, X)$ and $\text{Prog}_i(X)$ instead of $\text{TI}_{<_i}(s, X)$ and $\text{Prog}_{<_i}(X)$, respectively, for $i = 0, 1$. We also define

$$\alpha \subseteq_i X :\Leftrightarrow \forall \xi <_i \alpha. \xi \in X,$$

Then $\text{Prog}_i(X)$ is equivalent to $\forall \alpha \subseteq_i X. \alpha \in X$ and $\text{TI}_i(s, X)$ is equivalent to $\text{Prog}_i(X) \to s \subseteq_i X$. Corresponding to the defining operator for $\mathcal{A}cc$ we have the notion of Ω-*progressive*, i.e., progressive on the subclass of terms below Ω:

$$\text{Prog}_\Omega(X) :\Leftrightarrow \forall \alpha < \Omega. \alpha \subseteq X \to \alpha \in X.$$

By the inductive definition axioms for $\text{Acc}_{<_0}$ we obtain directly

(4.1) $\quad\quad\quad \mathcal{U}(\text{ID}_1) \vdash \alpha \in \text{Acc}_{<_0} \leftrightarrow \alpha \subseteq_0 \text{Acc}_{<_0},$
(4.2) $\quad\quad\quad \mathcal{U}(\text{ID}_1) \vdash \text{Prog}_0(\text{Acc}_{<_0}),$
(4.3) $\quad\quad\quad \mathcal{U}(\text{ID}_1) \vdash \text{Prog}_0(U) \to \text{Acc}_{<_0} \subseteq U,$

so we get

(4.4) $\quad\quad\quad \mathcal{U}(\text{ID}_1) \vdash \alpha \in \mathscr{A}cc \leftrightarrow (\alpha < \Omega \wedge \alpha \subseteq \mathscr{A}cc),$
(4.5) $\quad\quad\quad \mathcal{U}(\text{ID}_1) \vdash \text{Prog}_\Omega(\mathscr{A}cc),$
(4.6) $\quad\quad\quad \mathcal{U}(\text{ID}_1) \vdash \text{Prog}_\Omega(U) \to \mathscr{A}cc \subseteq U.$

LEMMA 4.3.1. *We have* $\mathcal{U}(\text{ID}_1) \vdash \text{Prog}_<(U) \to \text{Prog}_\Omega(U)$ *and thus also*

(4.7) $\quad\quad\quad \mathcal{U}(\text{ID}_1) \vdash \text{Prog}_<(U) \to \mathscr{A}cc \subseteq U,$
(4.8) $\quad\quad\quad \mathcal{U}(\text{ID}_1) \vdash \text{Prog}_1(U) \to \mathscr{A}cc \subseteq U.$

Proof. If $\alpha < \Omega$ and $\alpha \subseteq U$, then certainly $\alpha \subseteq U$, so $\alpha \in U$ by $\text{Prog}_<(U)$. Now (4.6) gives the second claim, which combined with Lemma 4.1.1 gives the third. □

LEMMA 4.3.2. *We have*

(4.9) $\quad\quad\quad \mathcal{U}(\text{ID}_1) \vdash (\forall \alpha \in \mathscr{A}cc.\, \alpha \subseteq U \to \alpha \in U) \to \mathscr{A}cc \subseteq U.$

Proof. Assuming the hypothesis, we need to prove $\forall \alpha \in \mathscr{A}cc.\, \alpha \in U$, which is equivalent to $\forall \alpha \in \mathscr{A}cc.\, \alpha \in \mathscr{A}cc \to \alpha \in U$. To this we apply the induction principle (4.6), yielding $\alpha < \Omega$ satisfying

$$\forall \beta < \alpha.\, \beta \in \mathscr{A}cc \to \beta \in U,$$

and the goal $\alpha \in \mathscr{A}cc \to \alpha \in U$. So assume $\alpha \in \mathscr{A}cc$. To prove $\alpha \in U$ we just apply our hypothesis, noting that any $\beta < \alpha$ is in $\mathscr{A}cc$ by (4.4). □

Note how (4.6), (4.7) and (4.8) allow us to prove something for an arbitrary element of $\mathscr{A}cc$ by a kind of induction. However, Lemma 4.3.2 is in some sense the strongest possible, so induction on $\mathscr{A}cc$ will refer to this principle.

LEMMA 4.3.3 ($\mathcal{U}(\mathrm{ID}_1)$). *The class Acc contains 0.*
Proof. This follows directly from (4.4). □

LEMMA 4.3.4 ($\mathcal{U}(\mathrm{ID}_1)$). *The class Acc is closed under ordinal addition.*
Proof. We want to prove $\forall \alpha \in Acc. \forall \beta \in Acc. \alpha + \beta \in Acc$. Let us apply induction on $\alpha \in Acc$, giving us $\alpha \in Acc$ with the inductive hypothesis

$$\forall \xi < \alpha. (\forall \eta \in Acc. \xi + \eta \in Acc)$$

and the goal $\forall \beta \in Acc. \alpha + \beta \in Acc$. Let us apply induction on $\beta \in Acc$, giving us $\beta \in Acc$ with the further inductive hypothesis

$$\forall \eta < \beta. \alpha + \eta \in Acc$$

and the goal $\alpha + \beta \in Acc$. This we prove by Ω-progressiveness of Acc, so our new goals are $\alpha + \beta < \Omega$ (this is clear) and $\alpha + \beta \subseteq Acc$. So take $\xi < \alpha + \beta$. If $\xi < \alpha$, we are done by assumption. Otherwise, $\xi = \alpha + \eta$ for some $\eta < \beta$, and we are done by the second inductive hypothesis. □

LEMMA 4.3.5 ($\mathcal{U}(\mathrm{ID}_1)$). *We have* $Acc \subseteq M$.
Proof. For $\alpha \in Acc$ we have $\alpha + 1 \subseteq Acc$, so by the properties of strongly critical components (Lemma 2.2.2) we get

$$SC(\alpha) \cap \Omega \subseteq \alpha + 1 \subseteq Acc.$$ □

LEMMA 4.3.6 ($\mathcal{U}(\mathrm{ID}_1)$). *The class Acc is closed under the $\bar{\varphi}$ function.*
Proof. We want to prove $\forall \alpha \in Acc. \forall \beta \in Acc. \bar{\varphi}_\alpha(\beta) \in Acc$. As in the proof of Lemma 4.3.4 we induct on Acc twice giving us $\alpha, \beta \in Acc$ satisfying

(4.10) $\qquad \forall \xi < \alpha. \forall \eta \in Acc. \bar{\varphi}_\xi(\eta) \in Acc,$
(4.11) $\qquad \forall \eta < \beta. \bar{\varphi}_\alpha(\eta) \in Acc,$

and leaving us with the new goal $\bar{\varphi}_\alpha(\beta) \in Acc$, which we prove using Ω-progressiveness, i.e., we need to prove $\bar{\varphi}_\alpha(\beta) < \Omega$ (which is clear) and

$$\forall \rho. \rho < \bar{\varphi}_\alpha(\beta) \to \rho \in Acc.$$

This in turn we prove by side induction on the term structure of ρ. Suppose $\rho < \bar{\varphi}_\alpha(\beta)$:

- $\rho = 0 \in \mathcal{A}cc$ by Lemma 4.3.3.
- If $\rho =_{\mathrm{NF}} \rho_1 + \cdots + \rho_n$, then each $\rho_i < \rho$ so $\rho_i \in \mathcal{A}cc$ by side induction hypothesis. Thus, $\rho \in \mathcal{A}cc$ by Lemma 4.3.4.
- If $\rho = \bar{\varphi}_\xi(\eta)$, then we have the following possibilities:
  1. $\xi < \alpha$ and $\eta < \bar{\varphi}_\alpha(\beta)$: Here we can use (4.10) by the side induction hypothesis for η.
  2. $\xi = \alpha$ and $\eta < \beta$: Here we can use (4.11) directly.
  3. $\xi > \alpha$ and $\rho \leq \beta$: Here we get $\rho \in \mathcal{A}cc$ from $\beta \in \mathcal{A}cc$ by (4.4).
- If $\rho \in \mathrm{SC}$, then $\rho \leq \mu$ for some $\mu \in \mathrm{SC}(\bar{\varphi}_\alpha(\beta)) = \mathrm{SC}(\alpha) \cup \mathrm{SC}(\beta) \subseteq \mathcal{A}cc$, so $\rho \in \mathcal{A}cc$ by (4.4). □

REMARK 4.3.7. SETZER (1998) described how in general, given an ordinal notation system, the class of accessible elements is closed under any part of the notation system that is built up or defined "from below", thus generalizing our Lemmata 4.3.4 and 4.3.6.

4.4 STEPPING UP

The following lemmata are similar to those of POHLERS (2009, Lemmata 9.6.17 and 18) concerning the lower bound of ID_1, and do not involve the unfolding machinery at all.[1]

LEMMA 4.4.1. $\mathcal{U}(\mathrm{ID}_1) \vdash \mathrm{TI}_1(\Omega + 1, U) \wedge \mathrm{K}(\Omega + 1) \subseteq \Omega + 1 \wedge \Omega + 1 \in \mathrm{M}$.
Proof. Since $\mathrm{K}(\Omega+1) = \mathrm{SC}(\Omega+1) = \emptyset$, we trivially have $\mathrm{K}(\Omega+1) \subseteq \Omega+1 \wedge \Omega+1 \in \mathrm{M}$. Now assume $\mathrm{Prog}_1(U)$, i.e., $\forall \alpha. \alpha \subseteq_1 U \to \alpha \in U$. Let $\alpha <_1 \Omega + 1$. We need to show $\alpha \in U$. It suffices to show $\forall \xi <_1 \alpha. \xi \in U$. So let $\xi \in \mathrm{M} \cap \Omega$. We know $\mathrm{SC}(\xi) \subseteq \mathcal{A}cc$, so $\xi \in \mathcal{A}cc$. Since we have $\mathrm{Prog}_1(U)$, we get $\xi \in U$ by (4.8), as desired. □

The idea for the following lemma is that the fundamental properties of Cantor Normal Forms hold for the $<_1$ relation.

---
[1] However, the proof of POHLERS (2009, Lemma 9.6.18) needs to be corrected, since the argument he gives for (4.13) below is wrong. POHLERS (1989, Lemma 29.13) is stated without proof, but the argument there in §15 in the case of PA carries over smoothly to the present setting.

LEMMA 4.4.2. *If*
$$\mathcal{U}(\mathrm{ID}_1) \vdash \mathrm{TI}_1(\alpha, U) \wedge K(\alpha) \subseteq \alpha \wedge \alpha \in M,$$
then
$$\mathcal{U}(\mathrm{ID}_1) \vdash \mathrm{TI}_1(\omega^\alpha, U) \wedge K(\omega^\alpha) \subseteq \omega^\alpha \wedge \omega^\alpha \in M.$$

Proof. We introduce a jump operator $\mathcal{J}$ by
$$\mathcal{J}(X) := \{\alpha \mid \alpha \in M \wedge \forall \xi \in M. \xi \subseteq_1 X \rightarrow \xi + \omega^\alpha \subseteq_1 X\}$$

for which we prove

(4.12) $$\mathcal{U}(\mathrm{ID}_1) \vdash \mathrm{Prog}_1(U) \rightarrow \mathrm{Prog}_1(\mathcal{J}(U)).$$

We reason internally and assume $\mathrm{Prog}_1(U)$ and take α such that $\alpha \subseteq_1 \mathcal{J}(U)$. To prove $\alpha \in \mathcal{J}(U)$ we further assume $\alpha \in M$ and take $\xi \in M$ such that $\xi \subseteq_1 U$ and let $\eta <_1 \xi + \omega^\alpha$. We need to show $\eta \in U$. For $\eta < \xi$ this follows using $\xi \subseteq_1 U$, and for $\eta = \xi$ it follows using additionally $\mathrm{Prog}_1(U)$. Otherwise, we shall find $\beta <_1 \alpha$ and a numeral n such that

(4.13) $$\xi < \eta < \xi + \omega^\beta \cdot n.$$

This follows by writing ξ and η in Cantor Normal Form using Lemma 2.2.4 and noting that the components are in M. Now we shall prove

$$\forall n. \xi + \omega^\beta \cdot n \subseteq_1 U$$

by ordinary induction on n. For $n = 0$ this is clear by the assumption $\xi \subseteq_1 U$, and the induction step follows since $\beta <_1 \alpha$ so $\beta \in \mathcal{J}(U)$.

Having established (4.12), we prove the lemma thus: Let $\alpha \in M$ and $K(\alpha) \subseteq \alpha$. Then $\omega^\alpha \in M$ and $K(\omega^\alpha) \subseteq \omega^\alpha$ by Lemma 2.2.3. Assuming $\mathcal{U}(\mathrm{ID}_1) \vdash \mathrm{TI}_1(\alpha, U)$, we get $\mathcal{U}(\mathrm{ID}_1) \vdash \mathrm{TI}_1(\alpha, \mathcal{J}(U))$. To prove $\mathcal{U}(\mathrm{ID}_1) \vdash \mathrm{TI}_1(\omega^\alpha, U)$, reason internally. From $\mathrm{Prog}_1(U)$ we get $\mathrm{Prog}_1(\mathcal{J}(U))$ by (4.12), so $\alpha \subseteq_1 \mathcal{J}(U)$, so $\alpha \in \mathcal{J}(U)$. Taking $\xi = 0$ then gives $\omega^\alpha \subseteq_1 U$, as desired. □

Since we have to be careful to preserve membership in M, the next lemma is not quite obvious. The idea is the same as for the previous lemma, though, namely to establish the fundamental properties of the binary Veblen function (and the normal forms thereof) for the $<_1$ relation.

LEMMA 4.4.3. *If*
$$\mathcal{U}(\mathrm{ID}_1) \vdash \mathrm{TI}_1(\alpha, U) \wedge \mathrm{K}(\alpha) \subseteq \alpha \wedge \alpha \in \mathrm{M},$$
then
$$\mathcal{U}(\mathrm{ID}_1) \vdash \mathrm{TI}_1(\varphi_\alpha 0, U) \wedge \mathrm{K}(\varphi_\alpha 0) \subseteq \varphi_\alpha 0 \wedge \varphi_\alpha 0 \in \mathrm{M}.$$

Since the proof is a little involved, we break it up into smaller parts. We follow the overall strategy of FEFERMAN AND STRAHM (2000, Theorem 3), but adapt the ideas from the lower bound proof of ID_1 (in particular Lemma 4.4.2 above).

The following is the only part that really exercises the unfolding machinery of $\mathcal{U}(\mathrm{ID}_1)$. Let $A(X, \alpha, x)$ be a formula with at most X, α, x free where we think of X as a binary predicate variable. We wish to define segments (in terms of the $<_1$-relation) of the A jump hierarchy starting with U, given set-theoretically by the transfinite recursion

$$Y_0 := \{ x \mid U(x) \},$$
$$Y_\alpha := \{ x \mid A(Y^\alpha, \alpha, x) \}$$

where $Y^\alpha := \{ (\beta, m) \mid \beta <_1 \alpha \wedge m \in Y_\beta \}$.

Using only the predicate axioms in group 4 in Section 1.4, we can find a term r_A such that $\mathcal{U}(\mathrm{ID}_1)$ proves

$$\Pi(X) \to \Pi(r_A(X, \alpha)) \wedge \forall x. x \in r_A(X, \alpha) \leftrightarrow A(X, \alpha, x).$$

Define a term hier_A by

$$\mathrm{hier}_A := \mathrm{fix}\Big(\lambda f, \alpha. \{ \text{if } \alpha = 0 \text{ then } \mathrm{pr}_U \text{ else } r_A\big(\mathrm{join}(f, (<_1 \alpha)), \alpha\big) \}\Big).$$

Here $(<_1 \alpha)$ is the predicate holding of those β such that $\beta <_1 \alpha$, and this is defined from the predicate operations in terms of the inductive definition, so we reflect the inductively defined predicate $\mathrm{Acc}_{<_0}$ as a predicate constant $\mathrm{acc}_{<_0}$. Note that we really need the *dependent join* operation here.

LEMMA 4.4.4 ($\mathcal{U}(\mathrm{ID}_1)$). $\mathrm{Prog}_1(\{ \alpha \mid \Pi(\mathrm{hier}_A(\alpha)) \})$.

Proof. Reasoning internally, assume α satisfies

(4.14) $$\forall \beta <_1 \alpha.\ \Pi(\mathrm{hier}_A(\alpha)).$$

We use Theorem 1.4.2 to prove $\Pi(\mathrm{hier}_A(\alpha))$, so we must prove

$$\Pi(\{\,\text{if } \alpha = 0 \text{ then } \mathrm{pr}_U \text{ else } r_A(\mathrm{join}(\mathrm{hier}_A, (<_1 \alpha)), \alpha)\,\}).$$

For $\alpha = 0$ this is clear; for $\alpha \neq 0$ we must prove $\Pi(\mathrm{join}(\mathrm{hier}_A, (<_1 \alpha)))$, but this follows from the join axiom using (4.14). □

LEMMA 4.4.5. *If*
$$\mathcal{U}(\mathrm{ID}_1) \vdash \mathrm{TI}_1(\alpha, U),$$
then
$$\mathcal{U}(\mathrm{ID}_1) \vdash \forall \beta <_1 \alpha.\ \Pi(\mathrm{hier}_A(\beta)).$$

Proof. This follows directly from the previous lemma by applying (SUBST) to the assumption. We substitute $\{\alpha \mid \Pi(\mathrm{hier}_A(\alpha))\}$ for U in $\mathrm{TI}_1(\alpha, U)$, which is allowed, because even though TI_1 references the inductive predicate, it does *not* reference its predicate type reflection. □

Now we turn to the specific operator that will enable the move from α to $\varphi_\alpha(0)$. First we define a different *jump operator*, called $\mathcal{K}$ (to distinguish it from the one from Lemma 4.4.2), defined in terms of the $<_1$ relation, as follows:

(4.15) $$\mathcal{K}(X) := \{\alpha \mid \alpha \in M \wedge \forall \xi \in M.\ \xi \subseteq_1 X \to \xi + \alpha \subseteq_1 X\,\}.$$

In analogy with (4.12) we have, only easier, that

(4.16) $$\mathcal{U}(\mathrm{ID}_1) \vdash \mathrm{Prog}_1(U) \to \mathrm{Prog}_1(\mathcal{K}(U)).$$

Let h and e be primitive recursive functions on ordinal notations such that
- $h(0) = e(0) = 0$,
- $h(\omega^\alpha) = 0$ and $e(\omega^\alpha) = \alpha$.
- if $\alpha = \omega^{\alpha_1} + \cdots + \omega^{\alpha_n}$ with $n > 1$ and $\alpha_1 \geq_1 \cdots \geq_1 \alpha_n$, then $h(\alpha) = \omega^{\alpha_1} + \cdots + \omega^{\alpha_{n-1}}$ and $e(\alpha) = \alpha_n$.

Note that $\mathrm{SC}(h(\alpha)) \cup \mathrm{SC}(e(\alpha)) \subseteq \mathrm{SC}(\alpha)$.

We define (following SCHÜTTE (1977, p. 185), which in turn builds on a suggestion by Feferman with an improvement by Schwichtenberg) an operator $\mathcal{A}$ as follows:

$$\mathcal{A}(X, \alpha, \eta) := \forall \xi. h(\alpha) \leq_1 \xi <_1 \alpha \to \varphi_{e(\alpha)}\eta \in \mathcal{K}(X_\xi)$$

LEMMA 4.4.6. $\mathcal{U}(\mathrm{ID}_1)$ *proves the implication: if* $\forall \beta <_1 \alpha. \Pi(\mathrm{hier}_\mathcal{A}(\beta))$, *then*

$$\forall \beta. 0 <_1 \beta <_1 \alpha \wedge (\forall \xi <_1 \beta. \mathrm{Prog}_1(\mathrm{hier}_\mathcal{A}(\xi))) \to \mathrm{Prog}_1(\mathrm{hier}_\mathcal{A}(\beta)).$$

Proof. Exactly as in SCHÜTTE (1977, Lemma 9 in §21.5), but included here for completeness.

We reason internally, and assume the hypothesis and take β and γ such that

(4.17) $\quad\quad\quad\quad\quad\quad\quad 0 <_1 \beta <_1 \alpha$

(4.18) $\quad\quad\quad\quad\quad\quad\quad \forall \eta <_1 \beta. \mathrm{Prog}_1(\mathrm{hier}_\mathcal{A}(\eta))$

(4.19) $\quad\quad\quad\quad\quad\quad\quad \gamma \subseteq_1 \mathrm{hier}_\mathcal{A}(\beta)$

We need to show $\gamma \in \mathrm{hier}_\mathcal{A}(\beta)$, that is, $\forall \xi. h(\beta) \leq_1 \xi <_1 \beta \to \varphi_{e(\beta)}\gamma \in \mathcal{K}(\mathrm{hier}_\mathcal{A}(\xi))$. By (4.16) it suffices to show $\forall \xi. h(\beta) \leq_1 \xi <_1 \beta \to \varphi_{e(\beta)}\gamma \subseteq_1 \mathcal{K}(\mathrm{hier}_\mathcal{A}(\xi))$, which is equivalent to $\forall \eta <_1 \varphi_{e(\beta)}\gamma. \forall \xi. h(\beta) \leq_1 \xi <_1 \beta \to \eta \in \mathcal{K}(\mathrm{hier}_\mathcal{A}(\xi))$. This in turn we prove by induction on the term structure of η. In each case we may assume $\eta <_1 \varphi_{e(\beta)}\gamma$, and fix ξ with $h(\beta) \leq_1 \xi <_1 \beta$:

- The case $\eta = 0$ is trivial.
- For $\eta = \Omega$ we get $\eta \in \mathcal{K}(\mathrm{hier}_\mathcal{A}(\xi))$ from (4.18), (4.16) and Lemma 4.4.1.
- The case $\eta = \eta_1 + \cdots + \eta_n$ is handled by a subsidiary induction on n, using $\eta_i <_1 \eta$ so by induction hypothesis, $\eta_i \in \mathcal{K}(\mathrm{hier}_\mathcal{A}(\xi))$.
- For $\eta = \varphi_{\eta_1}(\eta_2)$ there are three sub-cases:
  1. For $\eta_1 < e(\beta)$ put $\xi_1 := \xi + \omega^{\eta_1}$, so $0 <_1 \xi_1 <_1 \beta$. By induction hypothesis we get $\eta_2 \in \mathcal{K}(\mathrm{hier}_\mathcal{A}(\xi_1))$. Then $\eta_2 \in \mathrm{hier}_\mathcal{A}(\xi_1)$, that is,

     $$\forall \delta. h(\xi_1) \leq_1 \delta <_1 \xi_1 \to \varphi_{e(\xi_1)}(\eta_2) \in \mathcal{K}(\mathrm{hier}_\mathcal{A}(\delta)).$$

     Looking at the Cantor normal forms, we see that $h(\xi_1) \leq_1 \xi < \xi_1$ and $e(\xi_1) = \eta_1$, so $\eta \in \mathcal{K}(\mathrm{hier}_\mathcal{A}(\xi))$, as desired.
  2. If $\eta_1 = e(\beta)$, then $\eta_2 <_1 \gamma$, so $\eta_2 \in \mathrm{hier}_\mathcal{A}(\beta)$. Therefore, $\varphi_{e(\beta)}(\eta_2) \in \mathcal{K}(\mathrm{hier}_\mathcal{A}(\xi))$.

3. If $\eta_1 > e(\beta)$, then $\eta < \gamma$, so as in the previous case we get $\varphi_{e(\beta)}(\eta) \in \mathcal{K}(\text{hier}_\mathcal{A}(\xi))$. But $\eta \leq \varphi_{e(\beta)}(\eta)$, so $\eta \in \mathcal{K}(\text{hier}_\mathcal{A}(\xi))$, as desired.

- For $\eta = \psi(\eta_1)$ we have $\eta < \Omega$ since $\eta \in M$, so $\eta \in \text{Acc}$ and we get $\eta \in \mathcal{K}(\text{hier}_\mathcal{A}(\xi))$ from (4.18), (4.16) and (4.8). □

Proof of Lemma 4.4.3. We begin by assuming

$$\mathcal{U}(\text{ID}_1) \vdash \text{TI}_1(\alpha, U) \wedge K(\alpha) \subseteq \alpha \wedge \alpha \in M.$$

We use Lemma 4.4.2 to conclude

$$\mathcal{U}(\text{ID}_1) \vdash \text{TI}_1(\omega^\alpha, U) \wedge K(\omega^\alpha) \subseteq \omega^\alpha \wedge \omega^\alpha \in M.$$

Then also $\mathcal{U}(\text{ID}_1) \vdash \text{TI}_1(\omega^\alpha + 1, U)$, so by (SUBST), Lemmata 4.4.5 and 4.4.6 we get

(4.20) $\quad \mathcal{U}(\text{ID}_1) \vdash \text{TI}_1\big(\omega^\alpha + 1, \{\beta \mid \beta <_1 \omega^\alpha + 1 \to \text{Prog}_1(\text{hier}_\mathcal{A}(\beta))\}\big)$
(4.21) $\quad \mathcal{U}(\text{ID}_1) \vdash \forall \beta <_1 \omega^\alpha + 1. \Pi(\text{hier}_\mathcal{A}(\beta))$,
(4.22) $\quad \mathcal{U}(\text{ID}_1) \vdash \forall \beta. 0 <_1 \beta <_1 \omega^\alpha + 1 \wedge (\forall \xi <_1 \beta. \text{Prog}_1(\text{hier}_\mathcal{A}(\xi)))$
$\qquad \to \text{Prog}_1(\text{hier}_\mathcal{A}(\beta))$.

From now on we reason internally in $\mathcal{U}(\text{ID}_1)$. The parts $K(\varphi_\alpha(0)) \subseteq \varphi_\alpha(0)$ and $\varphi_\alpha(0) \in M$ are clear, so we need to prove $\text{TI}_1(\varphi_\alpha(0), U)$. So assume $\text{Prog}_1(U)$, or equivalently, $\text{Prog}_1(\text{hier}_\mathcal{A}(0))$. Together with (4.22) this yields

$$\forall \beta <_1 \omega^\alpha + 1. (\forall \xi <_1 \beta. \text{Prog}_1(\text{hier}_\mathcal{A}(\xi))) \to \text{Prog}_1(\text{hier}_\mathcal{A}(\beta)).$$

This is equivalent to $\text{Prog}_1(\{\beta \mid \beta <_1 \omega^\alpha + 1 \to \text{Prog}_1(\text{hier}_\mathcal{A}(\beta))\})$, so combined with (4.20) we get
$$\forall \beta <_1 \omega^\alpha + 1. \text{Prog}_1(\text{hier}_\mathcal{A}(\beta)).$$
In particular, $0 \in \text{hier}_\mathcal{A}(\omega^\alpha)$, so

$$0 \in r_\mathcal{A}(\text{join}(\text{hier}_\mathcal{A}, (<_1 \omega^\alpha)), \omega^\alpha).$$

Since $X_\xi = \text{hier}_\mathcal{A}(\xi)$ when $\xi <_1 \omega^\alpha$ and $X = \text{join}(\text{hier}_\mathcal{A}, (<_1 \omega^\alpha))$, this is in turn equivalent to

$$\forall \xi. h(\omega^\alpha) \leq_1 \xi <_1 \omega^\alpha \to \varphi_{e(\omega^\alpha)}(0) \in \mathcal{K}(\mathrm{hier}_{\mathcal{A}}(\xi)).$$

Since $h(\omega^\alpha) = 0$ and $e(\omega^\alpha) = \alpha$, we get

$$\forall \xi <_1 \omega^\alpha. \varphi_\alpha(0) \in \mathcal{K}(\mathrm{hier}_{\mathcal{A}}(\xi)).$$

Putting $\xi = 0$ we get $\varphi_\alpha(0) \in \mathcal{K}(U)$, which by the definition of $\mathcal{K}$ in (4.15) gives $\forall \xi \in M. \xi \subseteq_1 U \to \xi + \varphi_\alpha(0) \subseteq_1 U$. In particular, $\varphi_\alpha(0) \subseteq_1 U$, as desired. □

4.5 CONDENSATION

The following lemmata are proved like those of POHLERS (2009, Lemmata 9.6.15 and 16).
Define

$$\mathcal{A}cc_\Omega := \{\alpha \mid K(\alpha) \subseteq \alpha \land \alpha \in M \to \psi(\alpha) \in \mathcal{A}cc\}.$$

LEMMA 4.5.1. $\mathcal{U}(\mathrm{ID}_1) \vdash \mathrm{Prog}_1(\mathcal{A}cc_\Omega)$.

Proof. Reasoning internally, assume $\alpha \subseteq_1 \mathcal{A}cc_\Omega$, $K(\alpha) \subseteq \alpha$ and $\alpha \in M$. We prove $\psi(\alpha) \in \mathcal{A}cc$ using Ω-progressiveness of $\mathcal{A}cc$, so we need $\psi(\alpha) < \Omega$ (which is clear) and $\psi(\alpha) \subseteq \mathcal{A}cc$. We shall prove

$$\forall \rho. \rho < \psi(\alpha) \to \rho \in \mathcal{A}cc$$

by complete induction on the term structure of ρ. Suppose $\rho < \psi(\alpha)$:
- $\rho = 0 \in \mathcal{A}cc$ by Lemma 4.3.3.
- If $\rho =_{\mathrm{NF}} \rho_1 + \cdots + \rho_n$, then each $\rho_i < \rho$ so $\rho_i \in \mathcal{A}cc$ by induction hypothesis. Thus, $\rho \in \mathcal{A}cc$ by Lemma 4.3.4.
- If $\rho = \bar{\varphi}_\xi(\eta)$, then $\xi, \eta \leq \rho$, so $\xi, \eta \in \mathcal{A}cc$ by induction hypothesis. Thus, $\rho \in \mathcal{A}cc$ by Lemma 4.3.6.
- If $\rho =_{\mathrm{NF}} \psi(\rho_0)$, then $\rho_0 < \alpha$ and $K(\rho_0) \subseteq \rho_0$. To show that $\rho_0 \in M$ let $\xi \in \mathrm{SC}(\rho_0) \cap \Omega$. Then $\xi =_{\mathrm{NF}} \psi(\eta)$ for some $\eta \in K(\rho_0) \subseteq \alpha$. Thus $\xi = \psi(\eta) < \psi(\alpha)$, so by induction hypothesis for ξ we have $\xi \in \mathcal{A}cc$, as desired.

    Thus, $\rho_0 <_1 \alpha$, so $\rho_0 \in \mathcal{A}cc_\Omega$ which in turn implies that $\rho = \psi(\rho_0) \in \mathcal{A}cc$, since $\rho_0 \in M$. □

LEMMA 4.5.2 (Condensation). *If*

$$\mathcal{U}(\mathrm{ID}_1) \vdash \mathrm{TI}_1(\alpha, U) \wedge \mathrm{K}(\alpha) \subseteq \alpha \wedge \alpha \in \mathrm{M},$$

then

$$\mathcal{U}(\mathrm{ID}_1) \vdash \psi(\alpha) \in \mathcal{A}cc.$$

Proof. From the previous lemma and $\mathcal{U}(\mathrm{ID}_1) \vdash \mathrm{TI}_1(\alpha, \mathcal{A}cc_\Omega)$ we obtain $\mathcal{U}(\mathrm{ID}_1) \vdash \alpha \subseteq_1 \mathcal{A}cc_\Omega$. Using the previous lemma again we get $\mathcal{U}(\mathrm{ID}_1) \vdash \alpha \in \mathcal{A}cc_\Omega$ which together with $\mathrm{K}(\alpha) \subseteq \alpha$ and $\alpha \in \mathrm{M}$ implies $\mathcal{U}(\mathrm{ID}_1) \vdash \psi(\alpha) \in \mathcal{A}cc$, as desired. □

COROLLARY 4.5.3. *For every ordinal* $\alpha < \psi(\Gamma_{\Omega+1})$, $\mathcal{U}(\mathrm{ID}_1) \vdash \mathrm{TI}(\alpha, U)$.

Proof. This follows directly from Lemmata 4.4.1 to 4.4.3. □

Chapter 5

CONCLUSION

The main contribution of this thesis is the proof of the main theorem, which calculates the proof-theoretic ordinal of the unfolding of ID_1, namely $|\mathcal{U}(\mathrm{ID}_1)| = \psi(\Gamma_{\Omega+1})$. This thus settles a specific conjecture of Feferman.

The upper bound required a new combination of proof-theoretic techniques, specifically local predicativity via operator-controlled derivations and predicative asymmetric interpretation. Since operator-controlled derivations themselves are a form of impredicative asymmetric interpretation, this work opens the possibility for a unification of the techniques.

Let us now turn to other avenues of further work.

5.1 FURTHER WORK

Restricted join

For the unfolding of ID_1 it is necessary in the predicate structure to include a join operation that depends on a predicate to index the disjoint union, rather than just taking unions indexed by the natural numbers. It still remains to determine the exact strength of the unfolding of ID_1 when we only have the original, restricted join operator, join^r, axiomatized by:

$$(\forall y.\, \Pi(f(y))) \to \Pi(\mathrm{join}^r(f)) \wedge \forall x.\, x \in \mathrm{join}^r(f) \leftrightarrow \exists y, z.\, x = (y, z) \wedge z \in f(y).$$

The strength of the resulting system lies between $\psi(\varphi_2(\Omega + 1))$, that is, the collapse of the first fixed point of $\lambda \alpha.\, \varepsilon_\alpha$ after Ω, and $\psi(\Gamma_{\Omega+1})$. For the lower bound, this would follow from FEFERMAN AND STRAHM (2000, §6.1) together with the techniques in Chapter 4. I conjecture that the exact strength is in fact $\psi(\varphi_2(\Omega + 1))$, but at the moment I do not see how to establish the upper bound.

The unfolding of ID_ν

It would be very natural to study also the unfoldings the schematic systems of theories of (ν-times) iterated inductive definitions, ID_ν, where ν is an ordinal of a constructive ordinal notation system. These theories were introduced by FEFERMAN (1970a); the main reference for the basic setup and the original proof-theoretic analysis is the monograph of BUCHHOLZ, FEFERMAN, POHLERS, AND SIEG (1981). One of the main results is that $|\mathrm{ID}_\nu| = \psi(\varepsilon_{\Omega_\nu+1})$. The methods used in this thesis should apply without much difficulty to establish that $|\mathcal{U}(\mathrm{ID}_\nu)| = \psi(\Gamma_{\Omega_\nu+1})$, generalizing the result for ID_1.

Note that if we define
$$\mathrm{ID}_{<\nu} := \bigcup_{\xi<\nu} \mathrm{ID}_\xi,$$
then it would follow from such a result that $|\mathcal{U}(\mathrm{ID}_{<\nu})| = |\mathrm{ID}_{<\nu}|$ for limit ordinals ν. This doesn't contradict the fact that the unfoldings often prove the consistency of the starting system S, because that only happens when S has a finite number of schemata.

The general situation would be as in Table 5.1. The first line is from FEFERMAN AND STRAHM (2000), the second line from this thesis, and the other lines are conjectures.

Connections with iterated fixed points

It seems very likely that many unfolding systems can be interpreted using iterated fixed points. The rough idea is to model the predicate structure using the fixed points and let each use of the substitution rule correspond to one extra fixed point. Indeed,

T	$|\mathcal{U}_0(T)|$	$|\mathcal{U}(T)|$
NFA	ε_0	Γ_0
ID_1	$\psi(\varepsilon_{\Omega+1})$	$\psi(\Gamma_{\Omega+1})$
$\mathrm{ID}_{<\omega}$	$\psi(\Omega_\omega)$	$\psi(\Omega_\omega)$
ID_ω	$\psi(\varepsilon_{\Omega_\omega+1})$	$\psi(\Gamma_{\Omega_\omega+1})$
ID_ν	$\psi(\varepsilon_{\Omega_\nu+1})$	$\psi(\Gamma_{\Omega_\nu+1})$

Table 5.1: Strengths of theories; the ones below the line are conjectured.

the theory of finitely many fixed points, $\widehat{\text{ID}}_{<\omega}$, is equivalent to the unfolding of NFA, with both theories having proof-theoretic ordinal Γ_0 (for $\widehat{\text{ID}}_{<\omega}$ this was established by FEFERMAN (1982b)). Thus I conjecture that $\mathcal{U}(\text{ID}_1)$ is equivalent to a theory of finitely iterated fixed points on top of ID_1.

AVIGAD (1996) connected the first-order theory $\widehat{\text{ID}}_{<\omega}$ of finitely many iterated fixed points with the second order theory FP that formulates the existence of fixed points of arithmetical operators as a schema (notably with set parameters allowed):

(FP) $\quad\quad\quad\quad\quad\quad \exists X. \forall x.\, x \in X \leftrightarrow \mathcal{A}(X, x)$

He shows that these theories prove the same arithmetical sentences. This raises the question of whether the second order theory $\text{ID}_1^2 + (\text{FP})$ is equivalent to $\mathcal{U}(\text{ID}_1)$.

Parametrized arithmetical fixed points

It also seems quite natural to consider a theory that allows the formation of arithmetical inductive definitions relative to any previously introduced sets. A natural formalization of this idea is $\text{ID}_{<\omega}$, and we've conjectured in Section 5.1 that the strength of the unfolding of this theory is no stronger.

However, the formal unfolding system does not quite do the work it is supposed to do in this case, as we lose the possibility of applying the idea "form an inductive definition relative to already defined objects" in all generality. But I conjecture that a suitable formalization of unfolding for this idea will still be of the same strength as $\text{ID}_{<\omega}$.

Intuitionistic theories and type theories

As mentioned in Appendix B, extending the finite type structure for functionals on ordinals into the transfinite is a nontrivial matter. Something that has not been addressed in the work by ACZEL (1972) and HÖWEL (1977) is to what extent these structures qualify as extensions of the finitary standpoint in the sense of GÖDEL (1958). It seems much more promising to try to reconstruct the Aczel-Höwel functionals in a type theory that is uncontroversially constructive, such as type theory in the sense of MARTIN-LÖF (1975). Indeed, using the techniques of GRIFFOR AND RATHJEN (1994), PALMGREN (1992), and SETZER (1993), it should be possible to show for a Martin-Löf type theory with a base type for the natural numbers as well as a base type for the

49

second constructive tree ordinal class, and a predicative hierarchy of universes on top, both that

- the proof-theoretic strength is exactly that of $\mathcal{U}(\mathrm{ID}_1)$, which we have determined to be $\psi(\Gamma_{\Omega+1})$, and that
- suitable constructive versions of Höwel's H functionals are definable.

Indeed it seems puzzling, almost an accident of history that this theory has not already been analyzed in the literature. But working out this idea in detail must be left for another occasion.

We will note, though, that this would also help settle the conjecture of FEFERMAN (1991) that the schematic reflective closure of a suitable theory of constructive ordinals with a schematic principle of transfinite induction (similar to HOWARD 1972; TAKEUTI 1965) would be equivalent to some theory $\mathrm{ID}_{<\alpha}$ of iterated inductive definitions. Currently, I would conjecture the strength to be that of $\mathcal{U}(\mathrm{ID}_1)$, namely $\psi(\Gamma_{\Omega+1})$.

Unfolding of set theory

For the broader unfolding program, leaving the realm of inductive definitions, is the matter of unfolding set theory. As mentioned in Appendix A.3, Zermelo set theory with an open-ended schema of separation as well as Zermelo-Fraenkel set theory with the replacement schema, are prominent systems suitable for unfolding. Of course, they are too strong for a traditional proof-theoretic treatment, but perhaps they can be analyzed using ideas from the relativized ordinal analysis of RATHJEN (2014).

Appendix A

HISTORY OF UNFOLDINGS

A.1 MOTIVATION OF THE UNFOLDING PROGRAM

A very basic goal of the foundations of mathematics is the study of formal systems for the axiomatization of various mathematical concepts, particularly those that are rich.

It is well known since Gödel that any formal system encompassing already a weak system of arithmetic will be incomplete: the axioms will leave certain sentences undecided.

As a result, we are naturally led to search for stronger axioms. One instance of this is provided by Gödel's program for new axioms, that he advocated for most of his life, starting already in 1931 (FEFERMAN 1996). In particular, here are Gödel's remarks at the 1946 Princeton Bicentennial Conference:

> Let us consider, e.g., the concept of demonstrability. It is well known that, in whichever way you make it precise by means of a formalism, the contemplation of this very formalism gives rise to new axioms which are *exactly as evident and justified* as those with which you started, and this process of extension can be iterated into the transfinite. So there cannot exist any formalism which would embrace all these steps, but this does not exclude that all these steps (or at least all of them which give something new for the domain of propositions in which you are interested) could be described and collected together in some non-constructive way.
>
> (GÖDEL 1990, p. 151, emphasis added)

For the strongest foundational theories as of today, namely set theory with various large cardinal axioms, we would like to have a *uniform* justification for these axioms that we could use to find ever stronger axioms.

The unfoldings form just such a uniform procedure, first proposed by FEFERMAN (1996) and inspired in part by the emphasized part of the above quotation. However, there is an extensive background to his proposal.

A.2 EARLY FORERUNNERS OF UNFOLDINGS

Already Turing's thesis (TURING 1939) can be seen as an attempt at building stronger theories in a uniform way. Turing noted that the proof of the incompleteness theorems by itself gives the means of progressing from an incomplete system S_0 to a stronger system S_1 (by adjoining a consistency statement for S_0, $Con(S_0)$). By iterating this procedure we obtain an effective list of systems $\langle S_n \rangle_{n<\omega}$ from which we can form the effective union $S_\omega = \bigcup_{n<\omega} S_n$. The procedure can then be applied to S_ω to obtain $S_{\omega+1}$ and so on into the transfinite. But in order to ensure that all the systems are effectively presented, we need to use not ordinals per se, but rather *notations for constructive ordinals*, such as the system $\mathcal{O}$ of CHURCH AND KLEENE (1937). These encode the primitive recursive well-orderings on the natural numbers, and the least ordinal not expressible in this way is hence known as the Church–Kleene ordinal, ω_1^{CK}.

Turing considered the particular progressions where $S_0 = PA$, Peano Arithmetic, and $S_{\alpha+1} := S_\alpha \cup \{Con(S_\alpha)\}$. Since the consistency statements are Π_1^0, it makes sense to ask whether the union, $\bigcup_{\alpha<\omega_1^{CK}} S_\alpha$ is complete for Π_1^0-sentences of arithmetic.

Turing provided an affirmative answer to this question, given by a construction that to a particular Π_1^0-sentence associates a notation a that is in $\mathcal{O}$ just in case the sentence is true, and then is of height $\omega + 1$. This is of course disappointing; as Turing wrote:

> The theorem of completeness is also unexpected in that the ordinal formulæ used are all formulæ representing ω. This is contrary to our intentions in constructing Λ_P for instance; implicitly we had in mind large ordinals expressed in a simple manner. Here we have small ordinals expressed in a very complex and artificial way. (TURING 1939)

Here Λ_P is what Turing called an *ordinal logic*, that is, a family of systems indexed by $\mathcal{O}$. See the account of FEFERMAN (2006) for more details about the history and contents of Turing's thesis.

Technically, in order to form the limit stages of ordinal logics, we need to represent a recursively enumerable sequence of theories in a faithful way. This is not entirely trivial as can be seen by the work of FEFERMAN (1960) which treats in detail of the intensional aspects theories with infinitely many axioms. The results, both positive and negative, can be summarized as follows:

1. If T is a consistent extension of PA and $\tau(x)$ is a Σ_1^0 formula enumerating the

axioms of T in PA, then Con(τ) is not provable in T.
2. If T is *any* system whose axioms are enumerated by any formula $\tau(x)$ in PA, then T is interpretable in PA + Con(τ).
3. One can construct a bi-numeration[1] π^* of PA in PA such that PA $\vdash$ Con(π^*).
4. Moreover, for any reflexive[2] extension T of PA, one can construct a binumeration τ^* of T in T such that PA $\vdash$ Con(τ^*).

(See also FEFERMAN 1997.)

Ordinal logics were later renamed transfinite progressions of axiomatic theories in the study by FEFERMAN (1962). There, Feferman considered not only progressions based on adding consistency statements, but also the so-called *local* and *uniform reflection principles*.

Reflection principles

The local reflection principle is the schema,

(Rfn$_S$) $\qquad\qquad\qquad$ Prov$_S$($\ulcorner A \urcorner$) $\to$ A,

where A ranges over sentences in the language of S, and $\ulcorner A \urcorner$ denotes the Gödel number of A according to some acceptable coding (we will assume that all systems S under consideration contain enough machinery to allow such Gödel numbering).

We remark that (Rfn$_S$) allows us to derive Con(S) (take A to be a contradiction), so progressions based on the local reflection principle are stronger than those based on adding consistency statements. Thus, by Turing's result, starting with Peano arithmetic, we get completeness for Π_1^0-sentences, but Feferman showed, contrary to Turing's hope, that they are incomplete for Π_2^0-sentences (in the language of number theory, $\mathcal{L}^0$).

The uniform reflection principle states for $A(x)$ a formula in the language of S with only x free,

(RFN$_S$) $\qquad\qquad\qquad$ $\forall x.$ Prov$_S$($\ulcorner A(\dot{x}) \urcorner$) $\to$ $A(x)$.

---

[1] A bi-numeration is an enumeration each instance of which is decided by the target theory.

[2] A system is *reflexive* if it proves the consistency of each of its finite subsystems (represented as a finite disjunction of tests against the Gödel numbers of axioms).

For an acceptable system S, the uniform reflection principle, (RFN$_S$), is equivalent to a formalized ω-rule:

(RFN'$_S$) $\qquad\qquad (\forall x. \operatorname{Prov}_S(\ulcorner A(\dot{x})\urcorner)) \to \forall x. A(x).$

This observation is by FEFERMAN (1962, Theorem 2.19). Feferman showed that the progression starting from PA based on the uniform reflection principle is complete for all sentences in the language of arithmetic.

However, as shown by FEFERMAN AND SPECTOR (1962) these progressions are far from invariant (meaning notations for the same ordinals do not necessarily give equivalent theories), in that there are paths through $\mathcal{O}$ for which this progression is not even complete for Π^0_1-sentences.[3]

So a major problem with the notions based on transfinite progressions is the dependence on the choice of ordinal notations. In fact, a key observation is that the trick used to prove the above completeness theorem relies on choosing notations that cannot be seen to be notations on the basis of the starting system S_0.

Autonomous progressions

One possible answer to this worry is to restrict to *autonomous progressions*, where we are only allowed to progress to a system S_a when we have a proof in some previous system S_b that $a \in \mathcal{O}$. This idea is very fruitful and has been used to propose characterizations of *finitism* (KREISEL 1960, 1970) and *predicativity relative to the natural numbers* (FEFERMAN 1964, 1968).

A key part of these efforts is the attempt to capture *all principles of proof and ordinals which are implicit in given concepts* (KREISEL 1970). The reflective closure (and unfoldings), by way of contrast, tries to avoid reference to ordinal notations as a part of the basic concept, the idea being to try to explicate the implicit mathematical content of the concepts of S without any *prima facie* use of the notions of ordinal or well-ordering (FEFERMAN 1991).

Before we get to the unfoldings, however, I would like to mention the approaches based on *truth*, and before we can discuss either of these, it is necessary to digress in order to discuss *schematic theories*.

---

[3]See the work by FRANZÉN (2004a,b) for more details on transfinite progressions.

A.3 SCHEMATIC THEORIES

It is very common for mathematical theories to be formulated with one or more *schemata* as well as axioms. For example, consider the *induction schema* in Peano arithmetic, which is usually stated as

$$A(0) \wedge (\forall x.\, A(x) \to A(x')) \to \forall x.\, A(x),$$

where A ranges over formulæ in the language of number theory (with symbols for zero (0), successor ($'$), addition ($+$), multiplication ($\cdot$), and possibly further primitive recursive functions). This is the usual, modern formulation (apparently due to von Neumann around 1927; see the discussion and historical remarks by FEFERMAN 1991, §1.5,6). An earlier formulation would be

$$U(0) \wedge (\forall x.\, U(x) \to U(x')) \to \forall x.\, U(x),$$

where now U is a free unary predicate variable. In this formulation we have to adjoin to the logical calculus a rule of substitution

$$\frac{A(U)}{A(\{\, x \mid B(x)\,\})}$$

where A is any formula in the language, including predicate variables like U, and $B(x)$ is a formula with a distinguished free variable, x. The formula in the conclusion is obtained from A by replacing throughout any subformula $U(x)$ with $B(x)$ (replacing bound variables as necessary). The notation indicates that we think of $\{\, x \mid B(x)\,\}$ as representing the class of those x such that $B(x)$.

The chief advantage of this second formulation of schemata is that it more precisely reflects how we understand them: they are supposed to be valid for all definite classes, not just those we can formulate right now in the base language. It is this advantage that we can leverage to form the reflective closures and unfoldings. Let us therefore call schematic axioms in this second sense, *open-ended schemata*.

I should note (as already did FEFERMAN 1991, §1.5) that this is conceptually different from going to a second order axiom, say, for induction,

$$\forall X.\, 0 \in X \wedge (\forall x.\, x \in X \to x' \in X) \to \forall x.\, x \in X,$$

because this then brings in the full ontology of sets that relies on the comprehension scheme, either in the usual schematic form,

$$\exists X. \forall x. x \in X \leftrightarrow A(x),$$

or in the open-ended schematic form,

$$\exists X. \forall x. x \in X \leftrightarrow U(x).$$

Of course it is interesting to study this conception (both for second order theories, and for separation in Zermelo set theory), but it should be recognized that the additional ontology takes us far beyond what is warranted by the concepts implicit in an arbitrary starting system S.

A.4 REFLECTIVE CLOSURE

The next logical step after the approaches based on reflection comes from seeing that the reflection principles (Rfn$_S$) and (RFN$_S$) aim to express that whatever is derivable in a given system should be *true*. Thus, we can try to directly overlay a theory of truth on top of S in order to capture this idea more precisely.

Feferman (1991) first describes the so-called weak truth theory for any ordinary theory S, obtained by expanding the language by a truth predicate, T, and axiomatizing its application to statements of the original language, $\mathcal{L}$. The resulting theory is then a conservative extension of S that proves the Tarski truth schema ("convention T") for sentences A of the language of S, $T(\ulcorner A \urcorner) \leftrightarrow A$.

It is exactly at this point that the schematic theories come into play, because for these we now have the option of using all instances in the expanded language in the schemata of S. The resulting concept is the *ordinary truth theory* of a schematic theory (Feferman 1991, §2.5). A key result is the following theorem.

Theorem A.4.1. *Suppose* $S(U)$ *is a finitely axiomatized schematic theory extending schematic* PA, *in a language extending the usual language of number-theory,* $\mathcal{L}^0$. *Let* S *be the non-schematic version of* $S(U)$ *over* $\mathcal{L}^0$. *Then the ordinary truth theory of* $S(U)$ *proves*

$$\forall a. \text{Sent}_{\mathcal{L}^0}(a) \wedge \text{Prov}_S(a) \rightarrow T(a).$$

In particular, the ordinary truth theory proves the uniform reflection principle for S. Furthermore, this argument is formalizable in PA so the ordinary truth theory for PA proves $\mathrm{Con}(\mathrm{PA} + \mathrm{RFN}_{\mathrm{PA}})$.[4]

The ideas behind the self-applicative truth theories trace back at least to an influential paper by KRIPKE (1975). The formal systems I shall describe here date back to two lectures of Feferman for the Association of Symbolic Logic in 1979 and 1983, but this work was only published by FEFERMAN (1991) after several similar versions had appeared in the literature (cf. REINHARDT 1986 and CANTINI 1989; see also the introduction of FEFERMAN 1991).

The ordinary reflective closure $\mathrm{Ref}(S(U))$ is formulated in terms of partial truth and falsity predicates T and F and axioms for *self-reflecting truth* (KRIPKE 1975). These are self-applicable in the sense that they apply (codes of) sentences of the expanded language, $\mathcal{L}(T, F)$.

The schematic reflective closure $\mathrm{Ref}^*(S(U))$ goes beyond that and allows reasoning about *schematic truth* as well, i.e., asking which schemata ought to be accepted when one has accepted the axioms and schemata of S. This is achieved by expanding the language to $\mathcal{L}(U, T, F)$, adjoining axioms of self-applicable truth, now also applying to formulæ involving U, and then taking the substitution rule in the form,

$$\frac{A(U)}{A(\{\,x \mid B(x)\,\})}$$

where A is a formula of $\mathcal{L}(U)$, while $B(x)$ is any formula of $\mathcal{L}(U, T, F)$. We need the restriction because T and F can apply to (codes of) formulæ mentioning U.

FEFERMAN (1991) established the following proof-theoretic equivalences:

$$\mathrm{Ref}(\mathrm{PA}(U)) \equiv \mathrm{RA}_{<\varepsilon_0}$$
$$\mathrm{Ref}^*(\mathrm{PA}(U)) \equiv \mathrm{RA}_{<\Gamma_0}$$

JÄGER (1986, p. 123) sketched how the latter result can also be obtained by an embedding of reflective machinery in the theory KPi^0 of Kripke-Platek set theory with an inaccessible universe of sets but without foundation. Jäger used cut elimination with asymmetric interpretation to show that the strength of this theory is Γ_0; this

---

[4] The next step could be to study the iterations of the truth theories to obtain *ramified truth theories*. Starting with PA(U), these turn out to be intertranslatable with the usual theories of ramified analysis, RA_α, described by FEFERMAN (1964). Thus, we shall not pursue these further, but instead move on to approaches based on self-applicative or self-reflecting notions of truth.

technique is crucial for dealing with the unfolding systems as well as we saw in Chapter 3.

Relations to paradoxes

Partiality is a key idea in the self-applicable truth theories, so a small digression we can make at this point is to discuss its rôle in dealing with the usual paradoxes (e.g., Russell's paradox of sets $R = \{x \mid x \notin x\}$ and Tarski's paradox on the undefinability of truth).

FEFERMAN (1984) mentions three approaches to the paradoxes:
1. *Restriction of syntax.*
2. *Restriction of logic.*
3. *Restriction of basic principles.*

Various type theories take route 1, while Zermelo set theory opts for 3. It seems tempting to follow route 2, because it feels like statements such as $R \in R$ and the liar could be counted as undefined, somehow. That would make for an approach with three truth values, $\Omega = \{t, f, u\}$ but ordered such that u is below t and f. Thus, u represents not a definite truth value, but rather the absence of one. The connectives operate continuously on Ω via Kleene's strong three-valued logic, as shown in Fig. A.1 (for the binary connectives, read p down and q across).

However, as noted by FEFERMAN (1984), sustained reasoning directly using three-valued logic is almost impossible, so it is on this background that the reflective closures

p	$\neg p$
t	f
f	t
u	u

$p \wedge q$	t	f	u
t	t	f	u
f	f	f	f
u	u	f	u

$p \vee q$	t	f	u
t	t	t	t
f	t	f	u
u	t	u	u

$p \rightarrow q$	t	f	u
t	t	f	u
f	t	t	t
u	t	u	u

$p \leftrightarrow q$	t	f	u
t	t	f	u
f	f	t	u
u	u	u	u

Figure A.1: Kleene's strong three-valued logic

are built, using classical logic to reason about partial truth and falsity predicates.

Other truth theories

FRIEDMAN AND SHEARD (1987) introduced a series of axiomatized theories of self-applicable truth in classical logic and established the strength of a number of them. The remaining strengths were determined by LEIGH AND RATHJEN (2010).

CANTINI (1990) introduced a theory of truth, VF (named after van Fraassen's work on supervaluations) that he shows is equivalent to ID_1 for arithmetical sentences. This increase in strength comes about because of a desire to have all classical tautologies be true; thus truth is no longer required to be *grounded*.

A.5 THEORIES OF OPERATIONS

The problem with the various theories of truth from our perspective is that, going back to the quotation of Gödel mentioned in the beginning, the various truth axioms are hardly *exactly as evident and justified* as those of a system S. It is at this point that our insistence on the study of systems in classical first-order logic comes into play. Indeed, these systems are formulated in signatures consisting of both function and predicate symbols, so a very natural way to proceed will be to expand the range of operations on individuals (as expressed by function symbols) and also the range of operations on predicates (as expressed by the logical connectives) that the schematic theory can reason about.

There are a range of choices to be made in how to carry out this program. The first choice is how to represent abstract operations. The approach taken at first by FEFERMAN (1996) is to use recursive definability theory as pioneered in the thesis of PLATEK (1966) and developed as abstract recursion theory (FEFERMAN 1977; KECHRIS AND MOSCHOVAKIS 1977; MOSCHOVAKIS 1977, 1984). Here, we ask which operations of higher type should be considered recursively definable from a given collection of basic operations. It turns out that if we are just interested in the resulting base level operations, then we can restrict attention to objects of type level at most 2: PLATEK (1966) showed that total recursively definable operations can be defined using only objects of type level at most two, and PLOTKIN (1978) showed that once you have the partial continuous functionals of type 1 together with the type 2 continuous endofunctors on these, then you have the full partial continuous type structure. Thus, partial functions and functionals of type level 2 form the basic operational structure of the unfoldings as described in Appendix A.6.

The approach to unfoldings taken later by FEFERMAN AND STRAHM (2010) is to use abstract partial combinatory algebras to represent abstract operations. Combinatory algebra itself originated with SCHÖNFINKEL (1924) and the work of Curry (see CURRY AND FEYS 1958), who himself wanted to base logic on operational notions. Primarily in order to accommodate more examples, it is more convenient to use *partial* combinatory algebras, going back at least to work by FEFERMAN (1975) and Hyland (see HYLAND, JOHNSTONE, AND PITTS 1980).

A benefit of both of these approaches is that they confront a criticism of the approaches to reflection based on arithmetization of the logical machinery. Indeed, the arithmetized encodings, with their ensuing possibilities for non-canonical encodings with surprising properties, miss the in practice canonical nature of the logical machinery and the provability relation. In order to account for this, FEFERMAN (1989) (and earlier FEFERMAN 1982a), described a system for presenting logical systems in a finitary inductive style that more closely resembles the way we think about these informally. This style lends itself very well to formalization in the unfoldings, as we shall see below.

Another motivation for operations

For another way to see that it is necessary to invoke applicative theories (allowing self-application because of our type-free approach), consider that truth theories and similar theories allow us to define classes internally as formulæ with a distinguished free variable. If we let $a, b, f \in \text{Cl}$ denote classes in this sense, then we can try to declare that f is a function from a to b by saying it is a class of ordered pairs:

$$(f : a \to b) :\Leftrightarrow f \subseteq a \times b \wedge \forall x \in a. \exists! y \in b. \langle x, y \rangle \in f.$$

However, in this setting, we do not have that $a, b \in \text{Cl} \to \{ f \in \text{Cl} \mid f : a \to b \} \in \text{Cl}$. A counter-example is $\{0, 1\}^V$ where V is the universal class. This is a critical defect, function classes being fundamental for mathematical reasoning (FEFERMAN 1984, §13). But when operations are taken as primitive, function classes are recovered.

A.6 UNFOLDINGS

We thus finally get to the unfoldings themselves. In the previous section I mentioned the operational basis, either in terms of recursive definability theory, or in terms of partial combinatory algebra. This level is called the operational unfolding, usually denoted $\mathcal{U}_0(S)$ for a schematic system S.

To get the full unfolding, $\mathcal{U}(S)$, we add primitives for the atomic predicates of S and operations on predicates for conjunction, negation, universal quantification, as well as an operation for disjoint union of predicates. Sometimes we might wish to study the version without join; this is then called the intermediate unfolding system, $\mathcal{U}_1(S)$.

The main text described the version of unfolding that is based on a partial combinatory algebra (Section 1.4). Here we describe an earlier version based on recursive definability theory (cf. FEFERMAN (1996) and FEFERMAN AND STRAHM (2000)). In this version, the schematic unfolding $\mathcal{U}(S)$ contains three types of objects:

Typ 1 The basic types are ι (for individuals) and π_n (for n-ary predicates). We let κ, ν range over these types.

Typ 2 We have partial function types of the form $\vec{\iota} \rightharpoonup \iota$ and $\vec{\iota} \rightharpoonup \pi_n$. We let τ_0, σ_0 range over the former and τ, σ range over both.

Typ 3 $(\vec{\tau}_0, \vec{\iota} \rightharpoonup \iota)$ and $(\vec{\tau}, \vec{\kappa} \rightharpoonup \pi_n)$ are types of partial functionals.

The terms are generated as follows:

Tm 1 Individual variables of basic types κ. We use $x, y, z, \ldots$ as variables of type ι and $X^n, Y^n, Z^n, \ldots$ as variables of type π_n, omitting the superscript n when it can be inferred from the context.

Tm 2 Partial function variables $f, g, h, \ldots$ of the partial function types τ.

Tm 3 For each basic functional of S, a functional constant F_i of appropriate type, and for each basic predicate of S, a predicate constant P_i of appropriate type.

Tm 4 $\mathrm{Cond}(r, s) : (\vec{\tau}_0, \vec{\iota}, \iota, \iota \rightharpoonup \iota)$ for $r, s : (\vec{\tau}_0, \vec{\iota} \rightharpoonup \iota)$;
$\mathrm{Cond}(r, s) : (\vec{\tau}, \vec{\kappa}, \iota, \iota \rightharpoonup \pi_n)$ for $r, s : (\vec{\tau}, \vec{\kappa} \rightharpoonup \pi_n)$.

Tm 5 $r(\vec{s}, \vec{t}) : \iota$ for $r : (\vec{\tau}_0, \vec{\iota} \rightharpoonup \iota), \vec{s} : \vec{\tau}_0, \vec{t} : \vec{\iota}$;
$r(\vec{s}, \vec{t}) : \pi_n$ for $r : (\vec{\tau}, \vec{\kappa} \rightharpoonup \pi_n), \vec{s} : \vec{\tau}, \vec{t} : \vec{\kappa}$.

Tm 6 $(\lambda \vec{f}, \vec{x}. t) : (\vec{\tau}_0, \vec{\iota} \rightharpoonup \iota)$ for $\vec{f} : \vec{\tau}_0, \vec{x} : \vec{\iota}$, and $t : \iota$;
$(\lambda \vec{f}, \vec{x}. t) : (\vec{\tau}, \vec{\kappa} \rightharpoonup \pi_n)$ for $\vec{f} : \vec{\tau}, \vec{x} : \vec{\kappa}$, and $t : \pi_n$.

Tm 7 $\mathrm{LFP}(\lambda f, \vec{x}. t) : (\vec{\iota} \rightharpoonup \nu)$ for $f : \vec{\iota} \rightharpoonup \nu, \vec{x} : \vec{\iota}$, and $t : \nu$.

Furthermore, we have terms for operations on and to predicates:

Tm 8 $\mathrm{Eq} : \pi_2$.

Tm 9 $\mathrm{Pr}_U : \pi_1$ for the free predicate variable U.

Tm 10 $\mathrm{Inv}(s, t_1, \ldots, t_m) : \pi_n$ for $s : \pi_m, t_1, \ldots, t_m : \vec{\iota} \rightharpoonup \iota$ where the length of $\vec{\iota}$ is n.

Tm 11 $\mathrm{Neg}(t) : \pi_n$ for $t : \pi_n$.

Tm 12 $\mathrm{Conj}(s, t) : \pi_n$ for $s, t : \pi_n$.

Tm 13 $\mathrm{Un}(t) : \pi_n$ for $t : \pi_{n+1}$.

Tm 14 $\mathrm{Join}(t) : \pi_{n+1}$ for $t : \iota \rightharpoonup \pi_n$.

Tm 15 $\mathrm{DJoin}(t, s) : \pi_{n+1}$ for $t : \iota \rightharpoonup \pi_n$ and $s : \pi_1$.

The formulæ are given by:

Fm 1 The atomic formulæ are:

    (i) $r = s, r\downarrow$ and $U(r)$ for $r, s : \iota$, and U a free predicate symbol;

    (ii) $r = s, r\downarrow$ and $\vec{t} \in r$ for $r, s : \pi_n$, and $\vec{t} : \vec{\iota}$ of length n.

Fm 2 If A and B are formulæ, then so are also $\neg A$, $A \wedge B$, and $\forall x.\, A$.

The logic of $\mathcal{U}(S)$ is the logic of partial terms for the individual type ι, and classical quantifier free predicate logic for the other types.

    The axioms of $\mathcal{U}(S)$ are:

Ax 1 The axioms of S, including defining axioms for the F_i's and P_i's.

Ax 2 $\mathrm{Cond}(r, s)(\vec{f}, \vec{x}, y, y) \simeq r(\vec{f}, \vec{x}) \wedge$
$(y \neq z \to \mathrm{Cond}(r, s)(\vec{f}, \vec{x}, y, z) \simeq s(\vec{f}, \vec{x}))$.

Ax 3 $(\lambda \vec{f}, \vec{x}.\, t[\vec{f}, \vec{x}])(\vec{f}, \vec{x}) \simeq t[\vec{f}, \vec{x}]$.

Ax 4 For $s := \mathrm{LFP}(\lambda f, \vec{x}.\, t[f, \vec{x}])$ we take:

    (i) $s(\vec{x}) \simeq t[s, \vec{x}]$,

    (ii) $(\forall \vec{x}.\, f(\vec{x}) \simeq t[f, \vec{x}]) \to \forall \vec{x}.\, s(\vec{x}) \downarrow \to f(\vec{x}) = s(\vec{x})$.

Ax 5 $(\forall \vec{x}.\, \vec{x} \in X \leftrightarrow \vec{x} \in Y) \to X = Y$.

Ax 6 $\mathrm{Eq} \downarrow \wedge \forall x, y.\, (x, y) \in \mathrm{Eq} \leftrightarrow x = y$.

Ax 7 $\mathrm{Pr}_U \downarrow \wedge \forall x.\, x \in \mathrm{Pr}_U \leftrightarrow U(x)$.

Ax 8 $\mathrm{Inv}(X, f_1, \ldots, f_m) \downarrow \wedge$
$\forall \vec{x}.\, \vec{x} \in \mathrm{Inv}(X, f_1, \ldots, f_m) \leftrightarrow (f_1(\vec{x}), \ldots, f_m(\vec{x})) \in X$.

Ax 9 $\mathrm{Neg}(X) \downarrow \wedge \forall \vec{x}.\, \vec{x} \in \mathrm{Neg}(X) \leftrightarrow \vec{x} \notin X$.

Ax 10 $\mathrm{Conj}(X, Y) \downarrow \wedge \forall \vec{x}.\, \vec{x} \in \mathrm{Conj}(X, Y) \leftrightarrow \vec{x} \in X \wedge \vec{x} \in Y$.

Ax 11 $\mathrm{Un}(X) \downarrow \wedge \forall \vec{x}.\, \vec{x} \in \mathrm{Un}(X) \leftrightarrow \forall y.\, (\vec{x}, y) \in X$.

Ax 12 For $f : \iota \rightharpoonup \pi_n$ and $r : \pi_1$ we take

$$(\forall y.\, y \in r \to f(y)\downarrow) \to \mathrm{DJoin}(f, r)\downarrow \wedge$$
$$\forall \vec{x}, y.\, (\vec{x}, y) \in \mathrm{DJoin}(f, r) \leftrightarrow y \in r \wedge \vec{x} \in f(y).$$

Finally, we have the crucial predicate substitution rule:

$$\frac{A(U)}{A(\{\,x \mid B(x)\,\})}\ (\text{Subst})$$

Here, B is any formula, A is any formula not containing any subterms involving predicate types (as these can depend on U), and $A(\{\,x \mid B(x)\,\})$ is A where any subformula $U(t)$ is replaced by $\exists x.\, x \simeq t \wedge B(x)$ (to obey the strictness condition of the logic of partial terms).

Some subsystems are worth pointing out:
- For the operational unfolding, $\mathcal{U}_0(S)$, we leave out the predicate types, π_n, and any terms or axioms referring to them;
- For the intermediate unfolding $\mathcal{U}_1(S)$, we leave out the join axiom, Ax 12;
- For the restricted unfolding, we leave out the dependent join axiom, Ax 12, and replace it with the original join axiom that only branches over all individuals:
  Ax 12' For $f : \iota \to \pi_n$ we take

$$(\forall y.\, f(y)\!\downarrow) \to \text{Join}(f)\!\downarrow \wedge \forall \vec{x}, y.\, (\vec{x}, y) \in \text{Join}(f) \leftrightarrow \vec{x} \in f(y).$$

Previous results concerning unfoldings

FEFERMAN AND STRAHM (2000) studied the unfoldings of a basic schematic system for non-finitist arithmetic, S (using just 0, Sc and Pd), for which they established the following proof-theoretic equivalences:
- $\mathcal{U}_0(\text{NFA}) \equiv \text{PA}$,
- $\mathcal{U}_1(\text{NFA}) \equiv \text{RA}_{<\omega}$,
- $\mathcal{U}(\text{NFA}) \equiv \text{RA}_{<\Gamma_0}$,

where in each case the systems prove the same arithmetical sentences. In terms of proof-theoretic ordinals we get, respectively, ε_0, $\varphi_2(0)$, and Γ_0.

Following that, FEFERMAN AND STRAHM (2010) followed up with a study of the unfoldings of two different conceptions of finitist arithmetic, FA. In order to do so, they had to adapt the notion of unfolding to the setting of finitism where unbounded quantification is not considered meaningful. Instead one operates with schematic inference rules, and hence the relevant notion of substitution is that of substitution in such schematic inference rules.

For the most restricted conception, they proved that all three unfolding systems for finitist arithmetic, $\mathcal{U}_0(\text{FA})$, $\mathcal{U}_1(\text{FA})$, $\mathcal{U}(\text{FA})$, are proof-theoretically equivalent to Primitive Recursive Arithmetic, PRA. This supports Tait's analysis of finitism (TAIT 1968, 1981).

On the other hand, KREISEL (1965) had proposed that finitism should include transfinite recursion on recursive orderings for which one had proved the no descending sequences property. FEFERMAN AND STRAHM (2010) propose to model this with a finitistic version of the bar rule, BR, and they prove that all the corresponding unfolding systems, $\mathcal{U}_0(\text{FA} + \text{BR})$, $\mathcal{U}_1(\text{FA} + \text{BR})$, $\mathcal{U}(\text{FA} + \text{BR})$, are proof-theoretically equivalent to Peano Arithmetic, PA.

EBERHARD AND STRAHM (forthcoming) adapt unfoldings to a basic schematic system of feasible arithmetic, FEA. They study both the operational unfolding and two full unfoldings, one which incorporates a truth predicate, and prove for all of these that the provably total functions on binary words are precisely those computable in polynomial time.

Criticisms of foundations

Some people have criticized the usual approach to the foundations of mathematics via decidable proof systems (usually classical first-order logic or constructive type theories) on the basis that it fails to take into account the full lesson from the incompleteness phenomenon, and that to model mathematical reasoning, we need not only model inferences based on some fixed, however encompassing, axiomatization, but must also model the process through which mathematicians discover new axioms. See for instance CELLUCCI (1993) for an exposition of this view.

The unfoldings can be seen as a response to such criticisms in two ways: First, they explicitly acknowledge that common conceptions of mathematical structures employ *open-ended* schemata that are supposed to be valid for any accepted extension of the mathematical language or languages. Secondly, they propose particular ways in which the mathematical language can be extended in a natural way.

A.7 RELATIONS TO FREGE STRUCTURES

ACZEL (1980) studied Frege structures in order to give operational (or λ-calculi) models of the type theories of MARTIN-LÖF (1975). Frege structures are very closely related to the unfolding systems in that both incorporate operational machinery and some sort of reflection of predicate logic. But where the unfolding directly reflect predicates, Frege structures contain partial truth and falsity predicates T and F, and they are in this sense syntactically richer than unfoldings.

Unfolding in Frege Structures

To interpret the unfolding theories in truth theories, in particular theories for Frege structures, we can take predicates to be total propositional functions on the universe:

$$\Pi(x) := \forall y.\, T(\{x\}(y)) \lor F(\{x\}(y)),$$
$$y \in x := T(\{x\}(y))$$

This interpretation verifies the dependent version of the join axiom that is needed to obtain the natural strength of the unfolding of ID_1 (as we saw in Chapter 4). This further corroborates our feeling that that is a natural axiom.

Undefinability results

Aczel proves a number of undefinability results for Frege structures. First, there cannot be an internal definition of the collection of truths (echoing Tarski's theorem), and moreover, there cannot be an internal definition of the collection of propositions (since from such we could define the truths). And finally, the collection of sets is also not internally definable.

We take these negative results to show that in a Frege structure the notions of proposition, true proposition, and set, are open-ended notions. It also follows that the collection of subsets of a given set is also not internally definable.

A.8 RELATIONS TO EXPLICIT MATHEMATICS

The unfolding systems are also closely related to Feferman's systems of Explicit Mathematics (FEFERMAN 1975). These were originally introduced to provide an alternative foundation of Bishop's constructive mathematics that instead of relying on the constructive operational interpretation of logic, relies on operations to capture the explicit nature of Bishop's work.

Appendix B

ORDINAL ITERATION AND ACZEL'S ORDINAL

Bachmann's $H(1)$ is also called Aczel's ordinal, since it appeared in the study of ordinal iteration at transfinite types (ACZEL 1972). In this appendix, I will describe this work on iterations to explain the connections.

FEFERMAN (1970b) initiated the study of iteration functionals in the finite type structure over the countable ordinals, Ω. He introduced the notion of *repleteness* and showed that the critical process, Cr, at type level 2, preserves replete functions, and is thus *hereditarily replete*. The transfinite iteration functionals at each type level, It^{n+1}, were also shown to be hereditarily replete. It was conjectured that the least ordinal not expressible through these hereditarily replete primitive recursive functionals over Ω would be the Howard-Bachmann ordinal, $\psi(\varepsilon_{\Omega+1})$.[1] This was later verified by WEYHRAUCH (1976).

ACZEL (1972) proposed an extension of the finite type ordinal iterations using the following finite and transfinite types:

$$\Omega^{(0)} := \Omega;$$
$$\overline{\Omega}^{(\alpha)} := \prod_{\xi < \alpha} \Omega^{(\xi)}$$
$$\Omega^{(\alpha)} := \{ f : \overline{\Omega}^{(\alpha)} \to \Omega^{(0)} \}$$

For $x \in \overline{\Omega}^{(\alpha)}$ and $\beta < \alpha$ we define:

$$(x \,|_\beta) := \langle x_\xi \rangle_{\beta \leq \xi < \alpha}$$
$$(|_\beta\, x) := \langle x_\xi \rangle_{\xi < \beta}$$

Note that there is a natural isomorphism between functions f in $\Omega^{(\alpha)}$ and functions $f_\beta : \prod_{\beta \leq \xi < \alpha} \Omega^{(\xi)} \to \Omega^{(\beta)}$, for any $\beta < \alpha$, such that $f(x) = f_\beta(x \,|_\beta)(|_\beta\, x)$ for all

---

[1] See Chapter 2 for definitions of ordinal notations.

$x : \overline{\Omega}^{(\alpha)}$. In particular, for $\alpha = \beta + 1$ we have $f_\beta : \Omega^{(\beta)} \to \Omega^{(\beta)}$, thus generalizing the finite type structure.

For $f : \Omega^{(\alpha)}$, Aczel defines the transfinite iterations $f^\lambda : \Omega^{(\alpha)}$ by transfinite recursion:

$$f^1(x) := x$$
$$f^{\lambda+1}(x) := f^\lambda(\langle f^\lambda(x \mid_\beta) \rangle_{\beta < \alpha})$$
$$f^\lambda(x) := \sup_{\mu < \lambda} f^\mu(x).$$

However, as noted by HÖWEL (1977), this definition does not have the good properties that ACZEL (1972) needs, in particular Theorem 1.4 there is false. In private communication with Höwel, Aczel corrected his bounds by severely restricting the classes of functionals in the type structure. HÖWEL (1977) gave a more satisfactory solution by instead extending the type structure to include also types $\widetilde{\Omega}^{(\alpha)} = \alpha \to \Omega^{(\alpha)}$, that is, sequences of length equal to the type level in question.[2]

For $f : \widetilde{\Omega}^{(\alpha)}$ and $\beta < \alpha$ we write $f_\beta := f \beta$ and let $f^\beta : \prod_{\alpha \geq \xi > \beta} \Omega^{(\xi)} \to \Omega^{(\beta)}$ be the counterpart defined via $f^\beta (x \mid_\beta) (\mid_\beta x) := f_\beta x$.

For $f, g : \widetilde{\Omega}^{(\alpha)}$ Höwel defines $f \circ g : \widetilde{\Omega}^{(\alpha)}$ by:

$$(f \circ g)_\xi x := f^\xi \langle g^\delta (x \mid_\delta) \rangle_{\alpha > \delta \geq \xi} (\mid_\xi x).$$

Also, he defines $\text{Id}^\alpha : \widetilde{\Omega}^{(\alpha)}$ by

$$(\text{Id}^\alpha)_\xi x := x_\xi (\mid_\xi x).$$

Höwel shows that composition thus defined is associative with the Id functionals acting as left and right identities.

Having a suitable notion of composition, we can then define for $f : \widetilde{\Omega}^{(\alpha)}$, the transfinite iterations $f^\nu : \widetilde{\Omega}^{(\alpha)}$ by transfinite recursion:

$$f^0 := \text{Id}^\alpha,$$
$$f^{\nu+1} := f \circ f^\nu,$$
$$(f^\lambda)_\xi x := \sup_{\eta < \lambda} \{ (f^\eta)_\xi x \}.$$

---

[2] It should also be noted that Höwel's composition also works when the ground type is ω, where we can define iteration by diagonalization.

It can then be shown that $f^{\nu+\mu} = f^\nu \circ f^\mu$ and $(f^\nu)^\mu = f^{\nu\mu}$.

A single functional $f : \Omega^{(\alpha)}$ embeds as $\tilde{f} : \widetilde{\Omega}^{(\alpha)}$ by $\tilde{f}_\xi := f$ for $\xi < \alpha$. The given composition on the tilde types then allows us to recover the natural composition on successor types.

B.1 THE G FUNCTIONALS

The Grzegorczyk hierarchy can be generalized to Höwel's transfinite type structure, using the following bit of technicality:

For $z : \overline{\Omega}^{\alpha+1}$ decomposed as $z = y * x$ with $y : \Omega^{(\alpha)}$ and $z : \overline{\Omega}^{(\alpha)}$ we define

$$k(z) := \inf\{ \eta \in [\alpha', \alpha) \mid \forall \hat{\eta}. \eta \le \hat{\eta} < \alpha \to (\tilde{y}^\omega)_{\hat{\eta}} x = (\tilde{y}^\omega)_\eta x \},$$

where $\alpha =_{NF} \alpha' + \omega^\xi$ in Cantor normal form ($\alpha' = 0$ when $\alpha \in \mathbb{H}$). Then we write $\tilde{y}_k^\omega x$ short for $(\tilde{y}^\omega)_{k(y*x)} x$, and define inductively:

$$\begin{aligned}
G_0^1 x &:= x + 1, \\
G_0^{\alpha+1} y x &:= \tilde{y}_k^\omega x, \\
G_0^\lambda x &:= \sup\nolimits_{\xi<\lambda}\{ x_\xi (\vert_\xi x) \}, \\
G_{\gamma+1}^\alpha x &:= (\tilde{G}_\gamma^\alpha)_k^\omega x, \\
G_\lambda^\alpha x &:= \sup\nolimits_{\xi<\lambda} G_\xi^\alpha x.
\end{aligned}$$

Höwel proves that the limit of the autonomously expressible ordinals with the G functionals is Γ_0 by proving that

$$(\tilde{G}_0^{\omega^\beta})_1^{1+\gamma} \langle G_0^{1+\xi} \rangle_{\omega^\beta > \xi \ge 0} 0 = \varphi_\beta(\gamma),$$

thus recovering the Veblen hierarchy (Höwel 1977, p. 33).

B.2 THE H FUNCTIONALS

The G functionals only take us up to Γ_0, as they are based on ω-iteration only, and make no use of the critical process.

In his paper, Aczel (1972) describes an alternative family of functionals, there dubbed the I functionals. On p. 34, Höwel (1977) gives what he takes to be the analogue in his extended type structure:

$$H_0^\alpha := \mathrm{Id}_\alpha,$$
$$H_{\gamma+1}^\alpha x := \mathcal{E}\nu.\,(\tilde{H}_\gamma^\alpha)^{1+\nu}\,x,$$
$$H_\lambda^\alpha x := \sup_{\xi<\lambda} H_\xi^\alpha\,x.$$

Here, the construction $\mathcal{E}\nu.\,\theta(\nu)$ extracts the first fixed point a normal function θ, and in general we may set it equal to $\theta^\omega(0)$.

Höwel conjectures that the H functionals in finite types will express up to $H_0^\omega \langle H_0^\xi \rangle_{\omega > \xi \geq 0} = \psi(\varepsilon_{\Omega+1})$ and the full use will express ordinals up to Bachmann's $H(1)$, which in the next section we will prove is equal to the ordinal $\psi(\Gamma_{\Omega+1})$ described in Chapter 2.

B.3 COMPARISON OF ORDINALS

In this section we shall compare Bachmann's $H(1)$ with our $\psi(\Gamma_{\Omega+1})$. BACHMANN (1950) defines $H(1) := \tilde{\varphi}_{F_{\Omega_2+1}(1)}(1)$, where

1. $(F_\zeta)_{\zeta \leq \Omega_2+1}$ is a sequence of Ω_2-normal functions,[3] and
2. $(\tilde{\varphi}_\alpha)_{\alpha \leq F_{\Omega_2+1}(1)}$ is a sequence of Ω-normal functions,[4]

where for Bachmann, normal functions are defined for nonzero ordinals.

Bachmann defines (F_ζ) as the Veblen hierarchy of Ω_2-normal functions starting with $F_0(\xi) := \Omega^\xi$. Since the critical numbers of this function are precisely the ε-numbers after Ω, we see that $F_1(\xi) = \varphi_1(\Omega + \xi)$, and in general, $F_\zeta(\xi) = \varphi_\zeta(\Omega + \xi)$ for $0 < \zeta < \Omega_2$. Hence, following Bachmann's definition,

$$F_{\Omega_2}(\xi) := F_\xi(1) = \varphi_\xi(\Omega + 1),$$

so $F_{\Omega_2+1}(\xi) = \Gamma_{\Omega+\xi}$. In particular, $F_{\Omega_2+1}(1) = \Gamma_{\Omega+1}$.

To define the $\tilde{\varphi}$-functions, Bachmann assigns fundamental sequences $(\alpha[\xi])_{\xi \leq \tau_\alpha}$ (with the cofinality types τ_α being certain limit ordinals $\leq \Omega$) to limit ordinals $\alpha \leq \Gamma_{\Omega+1}$ such that $\alpha = \sup_{\xi < \tau_\alpha}$ and the maps $\xi \mapsto \alpha[\xi]$ are continuous and increasing.

The following definition is equivalent to Bachmann's:

DEFINITION B.3.1. For each limit $\alpha \leq \Gamma_{\Omega+1}$ define a fundamental sequence $(\alpha[\xi])_{\xi < \tau_\alpha}$:

---
[3] Bachmann uses ω_2 for our Ω_2.
[4] Bachmann calls these φ_η, but here we use $\tilde{\varphi}_\eta$ to distinguish them from the usual Veblen functions.

1. If $\alpha \leq \Omega$, then $\tau_\alpha := \alpha$ and $\alpha[\xi] := \xi$.
2. If $\Omega < \alpha =_{\text{NF}} \beta + \gamma$, then $\tau_\alpha := \tau_\gamma$ and $\alpha[\xi] := \beta + \gamma[\xi]$.
3. If $\Omega < \alpha =_{\text{NF}} \varphi_\beta(\gamma)$ ($\gamma \in \text{Lim}$), then $\tau_\alpha := \tau_\gamma$ and $\alpha[\xi] := \varphi_\beta(\gamma[\xi])$.
4. If $\Omega < \alpha =_{\text{NF}} \varphi_0(\gamma + 1) = \omega^{\gamma+1}$, then $\tau_\alpha := \omega$ and $\alpha[\xi] := \omega^\gamma \cdot \xi$.
5. If $\Omega < \alpha =_{\text{NF}} \varphi_{\beta+1}(0)$, then $\tau_\alpha := \omega$ and $\alpha[\xi] := \varphi_\beta^{(\xi)}(0)$.
6. If $\Omega < \alpha =_{\text{NF}} \varphi_{\beta+1}(\gamma + 1)$, then $\tau_\alpha := \omega$ and $\alpha[\xi] := \varphi_\beta^{(\xi)}(\varphi_{\beta+1}(\gamma) + 1)$.
7. If $\Omega < \alpha =_{\text{NF}} \varphi_\beta(0)$ ($\beta \in \text{Lim}$), then $\tau_\alpha := \tau_\beta$ and $\alpha[\xi] := \varphi_{\beta[\xi]}(0)$.
8. If $\Omega < \alpha =_{\text{NF}} \varphi_\beta(\gamma + 1)$ ($\beta \in \text{Lim}$), then $\tau_\alpha := \tau_\beta$ and $\alpha[\xi] := \varphi_{\beta[\xi]}(\varphi_\beta(\gamma) + 1)$.
9. If $\alpha = \Gamma_{\Omega+1}$, then $\tau_\alpha := \omega$, $\alpha[0] := \Omega + 1$, $\alpha[\xi + 1] := \varphi_{\alpha[\xi]}(0)$, $\alpha[\omega] := \alpha$.

To account for Bachmann's use of 1-indexed normal functions, we let $-1 + \xi$ be the unique ordinal η with $1 + \eta = \xi$ when $\xi \geq 1$. Then we define the functions $(\tilde{\varphi})_{\alpha \leq \Gamma_{\Omega+1}}$ as follows:

1. $\tilde{\varphi}_0(\xi) := \omega^\xi$.
2. $\tilde{\varphi}_{\alpha+1}(\xi) := \tilde{\varphi}'_\alpha(-1 + \xi)$.
3. If $\alpha \in \text{Lim}$ with $\tau_\alpha < \Omega$, then $\tilde{\varphi}_\alpha(\xi) := \text{en}_R(-1 + \xi)$ with $R = \bigcap_{\xi < \tau_\alpha} \text{Ran}(\tilde{\varphi}_{\alpha[\xi]})$.
4. If $\alpha \in \text{Lim}$ with $\tau_\alpha = \Omega$, then $\tilde{\varphi}_\alpha(\xi) := \tilde{\varphi}_{\alpha[\xi]}(1)$.

Instead of directly comparing the ψ-function with expressions in terms of the $(\tilde{\varphi}_\alpha)$-hierarchy we shall establish comparisons with the (θ_α)-hierarchy, which in turn lines up nicely with the Bachmann hierarchy.[5]

The idea of the (θ_α)-hierarchy originated with a proposal by Feferman (in unpublished work), and further work by ACZEL (1969), BRIDGE (1972, 1975), and WEYHRAUCH (1976) established the match up with the Bachmann hierarchy.[6] Here we use a version from notes of BUCHHOLZ (2011), dubbed $\hat{\theta}$.

Define simultaneously by recursion on α a family of sets $\text{Cl}_{\hat{\theta}}(\alpha, \beta)$ and Ω-normal functions $\hat{\theta}_\alpha$ by:

$\text{Cl}_{\hat{\theta}}(\alpha, \beta) :=$ the least set containing $\beta \cup \{0, \Omega\}$
and closed under $+$, the Veblen function $\lambda \xi \eta. \varphi_\xi(\eta)$,
and the restricted function $\lambda \xi < \alpha. \lambda \eta. \hat{\theta}_\xi(\eta)$,
$\hat{\theta}_\alpha := \text{en}\{\beta < \Omega \mid \beta \notin \text{Cl}_{\hat{\theta}}(\alpha, \beta) \land \alpha \in \text{Cl}_{\hat{\theta}}(\alpha, \beta)\}$.

---

[5]A more precise relation can be established using the results of SCHÜTTE (1986/87).
[6]See CROSSLEY AND BRIDGE KISTER (1986/87) for further details on these and other ordinal notation systems.

LEMMA B.3.2. *We have*
1. $\hat{\theta}_\alpha$ *is an Ω-normal function,*
2. $\mathrm{Cl}_{\hat{\theta}}(\alpha, \hat{\theta}_\alpha(\beta)) \cap \Omega = \hat{\theta}_\alpha(\beta)$.
3. *If* $\xi < \alpha$, $\xi \in \mathrm{Cl}_{\hat{\theta}}(\alpha, \hat{\theta}_\alpha(\beta))$, *and* $\eta < \hat{\theta}_\alpha(\beta)$, *then* $\hat{\theta}_\xi(\eta) < \hat{\theta}_\alpha(\beta)$.
4. $\hat{\theta}_{\alpha+1} = \hat{\theta}'_\alpha$.
5. *If* $\alpha \in \mathrm{Lim}$ *with* $\tau_\alpha < \Omega$, *then* $\mathrm{Ran}(\hat{\theta}_\alpha) = \bigcap_{\xi<\tau_\alpha} \mathrm{Ran}(\hat{\theta}_{\alpha[\xi]}) \setminus \{\tau_\alpha\}$.
6. *If* $\alpha \in \mathrm{Lim}$ *with* $\tau_\alpha = \Omega$, *then* $\mathrm{Ran}(\hat{\theta}_\alpha) = \{\beta \in \mathrm{Lim} \mid \beta \in \bigcap_{\xi<\beta} \mathrm{Ran}(\hat{\theta}_{\alpha[\xi]+1})$.

Accordingly, $H(1) = \hat{\theta}_{\Gamma_{\Omega+1}}(0)$.

In this section, we shall write Cl_ψ for the closure sets Cl defined in Section 2.1 in order to distinguish them from the $\mathrm{Cl}_{\hat{\theta}}$-sets.

LEMMA B.3.3. $\forall \alpha. \mathrm{Cl}_\psi(\alpha, 0) \subseteq \mathrm{Cl}_{\hat{\theta}}(\alpha, 0) \wedge \psi(\alpha) \le \hat{\theta}_\alpha(0)$.

Proof. By induction on α. The second conjunct follows from the first. For the first, induct on the build up of elements of $\mathrm{Cl}_\psi(\alpha, 0)$. The cases $0, \Omega, +,$ and φ are trivial. Suppose $\psi(\xi) \in \mathrm{Cl}_\psi(\alpha, 0)$ because $\xi \in \mathrm{Cl}_\psi(\alpha, 0) \cap \alpha$. By side induction hypothesis, $\xi \in \mathrm{Cl}_{\hat{\theta}}(\alpha, 0)$, and by main induction hypothesis, $\psi(\xi) \le \hat{\theta}_\xi(0) < \hat{\theta}_\alpha(0)$. Hence, $\psi(\xi) \in \mathrm{Cl}_{\hat{\theta}}(\alpha, 0)$, as desired. □

COROLLARY B.3.4. $\psi(\Gamma_{\Omega+1}) \le H(1)$.

Unfortunately, it's not so easy to get a bound in the other direction. The trouble is that to give an upper bound for $\hat{\theta}_\alpha(\beta)$ we need to have β sufficiently small; for example, we could imagine proving $\hat{\theta}_\alpha(\beta) \le \psi(g(\alpha, \beta))$ for some function g. But that is difficult to arrange, because for most g's $\lambda\beta. \psi(g(\alpha, \beta))$ would stabilize too quickly.

The solution is to use a variant of the ψ-function due to RATHJEN AND WEIERMANN (1993). We define closure sets Cl_ϑ and function values of ϑ by simultaneous recursion on α:

$$\mathrm{Cl}_\vartheta(\alpha, \beta) := \text{the least set containing } \beta \cup \{0, \Omega\}$$
$$\text{and closed under } +, \text{the Veblen function } \lambda\xi\eta.\, \varphi_\xi(\eta),$$
$$\text{and the restricted function } \lambda\xi < \alpha.\, \vartheta(\xi),$$
$$\vartheta(\alpha) := \min\{\xi < \Omega \mid \mathrm{Cl}_\vartheta(\alpha, \xi) \cap \Omega \subseteq \xi \wedge \alpha \in \mathrm{Cl}_\vartheta(\alpha, \xi)\}.$$

Then we can show by considering the corresponding notation systems that $\psi(\Gamma_{\Omega+1}) = \vartheta(\Gamma_{\Omega+1})$ (cf. RATHJEN AND WEIERMANN (1993, Corollary 3.2)).

LEMMA B.3.5. *For α we have:*
1. $\mathrm{Cl}_{\hat{\theta}}(\alpha, 0) \subseteq \mathrm{Cl}_{\vartheta}(\Omega^\alpha, 0) \wedge \hat{\theta}_\alpha(0) \leq \vartheta(\Omega^\alpha)$.
2. $\forall \beta. \mathrm{Cl}_{\hat{\theta}}(\alpha, \hat{\theta}_\alpha(\beta)+1) \subseteq \mathrm{Cl}_{\vartheta}(\Omega^\alpha \cdot (1+\beta+1), 0) \wedge \hat{\theta}_\alpha(\beta+1) \leq \vartheta(\Omega^\alpha \cdot (1+\beta+1))$.
3. $\forall \beta \in \mathrm{Lim}. \hat{\theta}_\alpha(\beta) \leq \vartheta(\Omega^\alpha \cdot (1+\beta))$.

The proof is by induction on α, and goes through smoothly since $\vartheta(\Omega^\alpha \cdot (1+\beta)) < \vartheta(\Omega^\alpha \cdot (1+\beta+1))$. This follows from RATHJEN AND WEIERMANN (1993, Lemma 1.2.7) since $\mathrm{SC}(\Omega^\alpha \cdot (1+\beta)) \cap \Omega \subseteq \mathrm{SC}(\Omega^\alpha \cdot (1+\beta+1)) \cap \Omega$.

COROLLARY B.3.6. $\hat{\theta}_{\Gamma_{\Omega+1}}(0) \leq \vartheta(\Gamma_{\Omega+1})$.

Thus, $H(1) = \hat{\theta}_{\Gamma_{\Omega+1}}(0) = \vartheta(\Gamma_{\Omega+1}) = \psi(\Gamma_{\Omega+1})$.

BIBLIOGRAPHY

I've tried to include links to all resources that are available online, through Digital Object Identifiers (DOI), or arXiv links. When this was not possible, I've included a link to Mathematical Reviews (MR) or Zentralblatt Math (ZMATH).

ACZEL, PETER (1969). "A new approach to the Bachmann method for describing countable ordinals. (Preliminary summary)". Unpublished.

— (1972). "Describing ordinals using functionals of transfinite type". In: *J. Symbolic Logic* 37 (1), pp. 35–47. DOI: 10.2307/2272543.

— (1980). "Frege structures and the notions of proposition, truth and set". In: *The Kleene Symposium (Proc. Sympos., Univ. Wisconsin, Madison, Wis., 1978)*. Vol. 101. Stud. Logic Foundations Math. Amsterdam: North-Holland, pp. 31–59. DOI: 10.1016/S0049-_237X(08)71252-_7.

AVIGAD, JEREMY (1996). "On the relationship between ATR_0 and $\widehat{ID}_{<\omega}$". In: *J. Symbolic Logic* 61.3, pp. 768–779. DOI: 10.2307/2275783.

BACHMANN, HEINZ (1950). "Die Normalfunktionen und das Problem der ausgezeichneten Folgen von Ordnungszahlen". In: *Vierteljschr. Naturforsch. Ges. Zürich* 95, pp. 115–147.

BEESON, MICHAEL J. (1986). "Proving programs and programming proofs". In: *Logic, methodology and philosophy of science, VII (Salzburg, 1983)*. Vol. 114. Stud. Logic Found. Math. Amsterdam: North-Holland, pp. 51–82. DOI: 10.1016/S0049-_237X(09)70684-_6.

BRIDGE, JANE (1972). "Some problems in mathematical logic. Systems of ordinal functions and ordinal notations". D. Phil. thesis. Oxford University.

— (1975). "A simplification of the Bachmann method for generating large countable ordinals". In: *J. Symbolic Logic* 40, pp. 171–185. DOI: 10.2307/2271898.

BUCHHOLZ, WILFRIED (1986). "A new system of proof-theoretic ordinal functions". In: *Ann. Pure Appl. Logic* 32.3, pp. 195–207. DOI: 10.1016/0168-_0072(86)90052-_7.

BUCHHOLZ, WILFRIED (1992). "A simplified version of local predicativity". In: *Proof theory (Leeds, 1990)*. Cambridge: Cambridge Univ. Press, pp. 115–147. DOI: 10.1017/CB09780511896262.006.

— (2001). "Finitary treatment of operator controlled derivations". In: *MLQ Math. Log. Q.* 47.3, pp. 363–396. DOI: 10.1002/1521-_3870(200108)47:3<_363::AID-_MALQ363>_3.0.CO;2-_P.

— (2011). "Comparison of $\hat{\theta}_\alpha$ and the Bachmann hierarchy below $\varepsilon_{\Omega+1}$". Unpublished notes.

BUCHHOLZ, WILFRIED, SOLOMON FEFERMAN, WOLFRAM POHLERS, AND WILFRIED SIEG (1981). *Iterated inductive definitions and subsystems of analysis: recent proof-theoretical studies*. Vol. 897. Lecture Notes in Mathematics. Berlin: Springer-Verlag, pp. v+383. DOI: 10.1007/BFb0091894.

CANTINI, ANDREA (1989). "Notes on formal theories of truth". In: *Z. Math. Logik Grundlag. Math.* 35.2, pp. 97–130. DOI: 10.1002/malq.19890350202.

— (1990). "A theory of formal truth arithmetically equivalent to ID_1". In: *J. Symbolic Logic* 55.1, pp. 244–259. DOI: 10.2307/2274965.

CELLUCCI, CARLO (1993). "From closed to open systems". In: *Philosophy of mathematics (Kirchberg am Wechsel, 1992)*. Vol. 20/. Schriftenreihe Wittgenstein-Ges. Vienna: Hölder-Pichler-Tempsky, pp. 206–220. URL: http://w3.uniroma1.it/cellucci/documents/Kirchberg.pdf.

CHURCH, ALONZO AND STEPHEN C. KLEENE (1937). "Formal definitions in the theory of ordinal numbers". In: *Fundamenta Mathematicae* 28 (1), pp. 11–21. URL: http://matwbn.icm.edu.pl/ksiazki/fm/fm28/fm2813.pdf.

CROSSLEY, JOHN N. AND JANE BRIDGE KISTER (1986/87). "Natural well-orderings". In: *Arch. Math. Logik Grundlag.* 26.1-2, pp. 57–76. DOI: 10.1007/BF02017491.

CURRY, HASKELL B. AND ROBERT FEYS (1958). *Combinatory logic. Vol. I.* Studies in logic and the foundations of mathematics. Amsterdam: North-Holland Publishing Co., pp. xvi+417. MR: 0094298.

EBERHARD, SEBASTIAN AND THOMAS STRAHM (forthcoming). "Unfolding feasible arithmetic and weak truth". In: *Unifying the Philosophy of Truth*. Ed. by D. ACHOURIOTI, H. GALINON, K. FUJIMOTO, AND J. MARTINEZ. Springer, 22 pp. URL: http://www.iam.unibe.ch/~strahm/download/pdf/unfolding_feasible.pdf.

FEFERMAN, SOLOMON (1960). "Arithmetization of metamathematics in a general setting". In: *Fundamenta Mathematicae* 49 (1), pp. 35–92. URL: http://matwbn.icm.edu.pl/ksiazki/fm/fm49/fm4915.pdf.

- (1962). "Transfinite recursive progressions of axiomatic theories". In: *J. Symbolic Logic* 27, pp. 259–316. DOI: 10.2307/2964649.
- (1964). "Systems of predicative analysis". In: *J. Symbolic Logic* 29, pp. 1–30. DOI: 10.2307/2269764.
- (1968). "Autonomous transfinite progressions and the extent of predicative mathematics". In: *Logic, Methodology and Philos. Sci. III (Proc. Third Internat. Congr., Amsterdam, 1967)*. Amsterdam: North-Holland, pp. 121–135. MR: 0252196.
- (1970a). "Formal theories for transfinite iterations of generalized inductive definitions and some subsystems of analysis". In: *Intuitionism and proof theory (Proc. Conf., Buffalo, N.Y., 1968)*. Amsterdam: North-Holland, pp. 303–326. DOI: 10.1016/S0049-237X(08)70761-4.
- (1970b). "Hereditarily replete functionals over the ordinals". In: *Intuitionism and Proof Theory (Proc. Conf., Buffalo, N. Y., 1968)*. Amsterdam: North-Holland, pp. 289–301. DOI: 10.1016/S0049-237X(08)70760-2.
- (1975). "A language and axioms for explicit mathematics". In: *Algebra and logic*. Vol. 450. Lecture Notes in Mathematics. Fourteenth Summer Res. Inst., Austral. Math. Soc., Monash Univ., Clayton, 1974: Springer, Berlin, pp. 87–139. DOI: 10.1007/BFb0062852.
- (1977). "Inductive schemata and recursively continuous functionals". In: *Logic Colloquium 76*. Oxford, 1976: North-Holland, Amsterdam, 373–392. Studies in Logic and Found. Math., Vol. 87. DOI: 10.1016/S0049-237X(09)70435-5.
- (1982a). "Inductively presented systems and the formalization of metamathematics". In: *Logic Colloquium '80 (Prague, 1980)*. Vol. 108. Stud. Logic Foundations Math. Amsterdam: North-Holland, pp. 95–128. DOI: 10.1016/S0049-237X(09)70506-3.
- (1982b). "Iterated inductive fixed-point theories: application to Hancock's conjecture". In: *Patras Logic Symposion (Patras, 1980)*. Vol. 109. Stud. Logic Foundations Math. Amsterdam: North-Holland, pp. 171–196.
- (1984). "Toward useful type-free theories. I". In: *J. Symbolic Logic* 49.1, pp. 75–111. DOI: 10.2307/2274093.
- (1989). "Finitary inductively presented logics". In: *Logic Colloquium '88 (Padova, 1988)*. Vol. 127. Stud. Logic Found. Math. Amsterdam: North-Holland, pp. 191–220. DOI: 10.1016/S0049-237X(08)70270-2.
- (1991). "Reflecting on incompleteness". In: *J. Symbolic Logic* 56.1, pp. 1–49. DOI: 10.2307/2274902.

FEFERMAN, SOLOMON (1996). "Gödel's program for new axioms: why, where, how and what?" In: *Gödel '96*. Vol. 6. Lecture Notes Logic. Brno, 1996: Springer, Berlin, pp. 3–22. Project Euclid: 1235417011.

— (1997). "My route to arithmetization". In: *Theoria* 63.3. The arithmetization of metamathematics, pp. 168–181. DOI: 10.1111/j.1755-_2567.1997.tb00746.x.

— (2006). "Turing's thesis". In: *Notices Amer. Math. Soc.* 53.10, pp. 1200–1206. URL: http://www.ams.org/notices/200610/fea-_feferman.pdf.

FEFERMAN, SOLOMON AND CLIFFORD SPECTOR (1962). "Incompleteness along paths in progressions of theories". In: *J. Symbolic Logic* 27, pp. 383–390. URL: http://www.jstor.org/stable/2964544.

FEFERMAN, SOLOMON AND THOMAS STRAHM (2000). "The unfolding of non-finitist arithmetic". In: *Annals of Pure and Applied Logic* 104.1-3, pp. 75–96. DOI: 10.1016/S0168-_0072(00)00008-_7.

— (2010). "Unfolding finitist arithmetic". In: *Rev. Symb. Log.* 3.4, pp. 665–689. DOI: 10.1017/S1755020310000183.

FRANZÉN, TORKEL (2004a). *Inexhaustibility*. Vol. 16. Lecture Notes in Logic. A non-exhaustive treatment. Urbana, IL: Association for Symbolic Logic, pp. xii+251. MR: 2269896.

— (2004b). "Transfinite progressions: a second look at completeness". In: *Bull. Symbolic Logic* 10.3, pp. 367–389. DOI: 10.2178/bsl/1102022662.

FRIEDMAN, HARVEY AND MICHAEL SHEARD (1987). "An axiomatic approach to self-referential truth". In: *Ann. Pure Appl. Logic* 33.1, pp. 1–21. DOI: 10.1016/0168-_0072(87)90073-_X.

GERBER, HARVEY (1970). "Brouwer's bar theorem and a system of ordinal notations". In: *Intuitionism and Proof Theory (Proc. Conf., Buffalo, N.Y., 1968)*. Amsterdam: North-Holland, pp. 327–338. DOI: 10.1016/S0049-_237X(08)70762-_6.

GÖDEL, KURT (1958). "Über eine bisher noch nicht benützte Erweiterung des finiten Standpunktes". In: *Dialectica* 12, pp. 280–287.

— (1990). *Collected works. Vol. II*. Publications 1938–1974, Edited and with a preface by Solomon Feferman. New York: The Clarendon Press Oxford University Press, pp. xviii+407. MR: 1032517.

GRIFFOR, EDWARD AND MICHAEL RATHJEN (1994). "The strength of some Martin-Löf type theories". In: *Arch. Math. Logic* 33.5, pp. 347–385. DOI: 10.1007/BF01278464.

HOWARD, WILLIAM A. (1972). "A system of abstract constructive ordinals". In: *J. Symbolic Logic* 37, pp. 355–374. DOI: 10.2307/2272979.

HÖWEL, KARL-ADOLF (1977). "Iteration von Funktionalen transfiniter Typen". PhD thesis. Westfälische Wilhelms-Universität Münster.

HYLAND, J. M. E., P. T. JOHNSTONE, AND A. M. PITTS (1980). "Tripos theory". In: *Math. Proc. Cambridge Philos. Soc.* 88.2, pp. 205–231. DOI: 10.1017/S0305004100057534.

JÄGER, GERHARD (1986). *Theories for admissible sets: a unifying approach to proof theory*. Vol. 2. Studies in Proof Theory. Lecture Notes. Bibliopolis, Nables, pp. iv+167. MR: 881218.

KECHRIS, ALEXANDER S. AND YIANNIS N. MOSCHOVAKIS (1977). "Recursion in Higher Types". In: *Handbook of Mathematical Logic*. Ed. by Jon BARWISE. Vol. 90. Studies in Logic and the Foundations of Mathematics. Elsevier, pp. 681–737. DOI: 10.1016/S0049-_237X(08)71119-_4.

KLEENE, STEPHEN C. (1945). "On the interpretation of intuitionistic number theory". In: *J. Symbolic Logic* 10, pp. 109–124. URL: http://www.jstor.org/stable/2269016.

KLEENE, STEPHEN C. AND RICHARD E. VESLEY (1965). *The foundations of intuitionistic mathematics, especially in relation to recursive functions*. Amsterdam: North–Holland Publishing Co., pp. viii+206. MR: 0176922.

KREISEL, GEORG (1960). "Ordinal logics and the characterization of informal concepts of proof". In: *Proc. Internat. Congress Math. 1958*. New York: Cambridge Univ. Press, pp. 289–299. URL: http://www.mathunion.org/ICM/ICM1958/Main/icm1958.0289.0299.ocr.pdf.

— (1965). "Mathematical logic". In: *Lectures on Modern Mathematics, Vol. III*. New York: Wiley, pp. 95–195. MR: 0177866.

— (1970). "Principles of proof and ordinals implicit in given concepts". In: *Intuitionism and Proof Theory (Proc. Conf., Buffalo, N. Y., 1968)*. Amsterdam: North-Holland, pp. 489–516. DOI: 10.1016/S0049-_237X(08)70773-_0.

KRIPKE, SAUL (1975). "Outline of a Theory of Truth". In: *The Journal of Philosophy* 72 (19). Seventy-Second Annual Meeting American Philosophical Association, Eastern Division (Nov. 6, 1975), pp. 690–716. URL: http://www.jstor.org/stable/2024634.

LEIGH, GRAHAM E. AND MICHAEL RATHJEN (2010). "An ordinal analysis for theories of self-referential truth". In: *Arch. Math. Logic* 49.2, pp. 213–247. DOI: 10.1007/s00153-_009-_0170-_2.

MARTIN-LÖF, PER (1975). "An intuitionistic theory of types: predicative part". In: *Logic Colloquium '73 (Bristol, 1973)*. Amsterdam: North-Holland, 73–118. Studies in Logic and the Foundations of Mathematics, Vol. 80. MR: 0387009.

MILLER, LARRY W. (1976). "Normal functions and constructive ordinal notations". In: *J. Symbolic Logic* 41.2, pp. 439–459. DOI: 10.2307/2272243.

MOSCHOVAKIS, YIANNIS N. (1977). "On the basic notions in the theory of induction". In: *Logic, Foundations of Mathematics, and Computability Theory. Part One of the Proceedings of the Fifth International Congress of Logic, Methodology and Philosophy of Science, London, Ontario, Canada, 1975*. Ed. by Robert. E BUTTS AND JAAKKO HINTIKKA. Vol. 9. The University of Western Ontario Series in Philosophy of Science. Springer, pp. 207–236. DOI: 10.1007/978-_94-_010-_1138-_9_12.

— (1984). "Abstract recursion as a foundation for the theory of algorithms". In: *Computation and proof theory (Aachen, 1983)*. Vol. 1104. Lecture Notes in Math. Berlin: Springer, pp. 289–364. DOI: 10.1007/BFb0099491.

PALMGREN, ERIK (1992). "Type-theoretic interpretation of iterated, strictly positive inductive definitions". In: *Arch. Math. Logic* 32.2, pp. 75–99. DOI: 10.1007/BF01269951.

PLATEK, RICHARD A. (1966). "Foundation of Recursion Theory". PhD thesis. Stanford University. MR: 2615453.

PLOTKIN, GORDON (1978). "$\mathbb{T}^\omega$ as a universal domain". In: *J. Comput. System Sci.* 17.2, pp. 209–236. DOI: 10.1016/0022-_0000(78)90006-_5.

POHLERS, WOLFRAM (1989). *Proof theory*. Vol. 1407. Lecture Notes in Mathematics. An introduction. Berlin: Springer-Verlag, pp. vi+213. DOI: 10.1007/978-_3-_540-_46825-_7.

— (1998). "Subsystems of set theory and second order number theory". In: *Handbook of proof theory*. Vol. 137. Stud. Logic Found. Math. Amsterdam: North-Holland, pp. 209–335. DOI: 10.1016/S0049-_237X(98)80019-_0.

— (2009). *Proof theory*. Universitext. The first step into impredicativity. Berlin: Springer-Verlag, pp. xiv+370. DOI: 10.1007/978-_3-_540-_69319-_2.

RATHJEN, MICHAEL (2014). "Relativized ordinal analysis: The case of Power Kripke–Platek set theory". In: *Ann. Pure Appl. Logic* 165.1, pp. 316–339. DOI: 10.1016/j.apal.2013.07.016.

RATHJEN, MICHAEL AND ANDREAS WEIERMANN (1993). "Proof-theoretic investigations on Kruskal's theorem". In: *Ann. Pure Appl. Logic* 60.1, pp. 49–88. DOI: 10.1016/0168-_0072(93)90192-_G.

REINHARDT, WILLIAM N. (1986). "Some remarks on extending and interpreting theories with a partial predicate for truth". In: *J. Philos. Logic* 15.2, pp. 219–251. DOI: 10.1007/BF00305492.

SCHÖNFINKEL, MOSES (1924). "Über die Bausteine der mathematischen Logik". In: *Math. Ann.* 92.3-4, pp. 305–316. DOI: 10.1007/BF01448013.

SCHÜTTE, KURT (1977). *Proof theory*. Translated from the revised German edition by J. N. Crossley, Grundlehren der Mathematischen Wissenschaften, Band 225. Berlin: Springer-Verlag, pp. xii+299. MR: 0505313.

— (1986/87). "Majorisierungsrelationen und Fundamentalfolgen eines Ordinalzahlensystems von G. Jäger". In: *Arch. Math. Logik Grundlag.* 26.1-2, pp. 29–55. DOI: 10.1007/BF02017490.

SETZER, ANTON (1993). "Proof theoretical strength of Martin-Löf Type Theory with W-type and one universe". PhD thesis. Ludwig-Maximilians-Universität München. URL: http://www.cs.swan.ac.uk/~csetzer/articles/weor0.pdf.

— (1998). "An introduction to well-ordering proofs in Martin-Löf's type theory". In: *Twenty-five years of constructive type theory (Venice, 1995)*. Vol. 36. Oxford Logic Guides. New York: Oxford Univ. Press, pp. 245–263. URL: http://www.cs.swan.ac.uk/~csetzer/articles/venedig.ps.

STRAHM, THOMAS (2000). "The non-constructive μ operator, fixed point theories with ordinals, and the bar rule". In: *Proceedings of the Workshop on Proof Theory and Complexity, PTAC'98 (Aarhus)*. Vol. 104. 1-3, pp. 305–324. DOI: 10.1016/S0168-_0072(00)00016-_6.

TAIT, WILLIAM W. (1968). "Constructive reasoning". In: *Logic, Methodology and Philos. Sci. III (Proc. Third Internat. Congr., Amsterdam, 1967)*. Amsterdam: North-Holland, pp. 185–199. MR: 0256877.

— (1981). "Finitism". In: *The Journal of Philosophy* 78 (9), pp. 524–546. URL: http://www.jstor.org/stable/2026089.

TAKEUTI, GAISI (1965). "A formalization of the theory of ordinal numbers". In: *J. Symbolic Logic* 30, pp. 295–317. DOI: 10.2307/2269620.

TUPAILO, SERGEI (2000). "Finitary reductions for local predicativity. I. Recursively regular ordinals". In: *Logic Colloquium '98 (Prague)*. Vol. 13. Lect. Notes Log. Urbana, IL: Assoc. Symbol. Logic, pp. 465–499. URL: http://www.cs.ioc.ee/~sergei/Mypapers/lp.final.ps.

TURING, ALAN M. (1939). "Systems of Logic Based on Ordinals". In: *Proc. London Math. Soc.* 2nd ser. 45.1, pp. 161–228. DOI: 10.1112/plms/s2-_45.1.161.

WEYHRAUCH, RICHARD (1976). "Relations between some hierarchies of ordinal functions and functionals". Completed and circulated in 1972. PhD thesis. Stanford University.

SYMBOL INDEX

$\mathrm{Cl}(\alpha, \beta)$ α-closure of β 12
$\mathrm{Cl}_\Omega(\alpha) = \mathrm{Cl}(\alpha, \psi(\alpha))$ 13

$\mathrm{dg}(F)$ degree of $\mathcal{L}^{1,r}_\infty$-formula F 30

$\mathcal{H}$ operator $\mathcal{H}\colon \mathrm{Pow}(\mathrm{On}) \to \mathrm{Pow}(\mathrm{On})$ 24
$\mathcal{H}_\alpha$ α-th iterated acceptable operator 24
$\mathcal{H}[X]$ the operator $\lambda Y.\, H(X \cup Y)$ 24

$\mathcal{L}^0$ the language of PA 53
$\mathcal{L}^1$ the language of ID_1 7
$\mathcal{L}^1_\infty$ the infinitary language for the upper bound for $\mathcal{U}(\mathrm{ID}_1)$ 20
$\mathcal{L}^{1,c}_\infty$ fragment of $\mathcal{L}^1_\infty$ suitable for T 21
$\mathcal{L}^{1,r}_\infty$ fragment of $\mathcal{L}^1_\infty$ suitable for T^r 21
$\mathcal{L}^{1,rc}_\infty := \mathcal{L}^{1,r}_\infty \cap \mathcal{L}^{1,c}_\infty$ 21
$\mathcal{L}^1_{\mathrm{On}}$ the language of $(\mathrm{ID}_1)^+_{\mathrm{On}}$ 18

$\mathrm{par}(E)$ the ordinal parameters in an $\mathcal{L}^1_\infty$-expression E 22
$\psi(\alpha)$ the αth collapse of Ω 12

$\mathrm{rk}(A)$ rank of $\mathcal{L}^{1,r}_\infty$-formula A 22

$t_F(G)$ term giving characteristic subformula G of F 23

$\mathcal{U}(S)$ Full unfolding of schematic system S 61
$\mathcal{U}_0(S)$ Operational unfolding of schematic system S 61
$\mathcal{U}_1(S)$ Intermediate unfolding of schematic system S 61

INDEX

accessible part, 35, 36
asymmetric interpretation, 31, 32, 34, 58

calculus
    derived, 29
    restricted, 24
Cantor normal form, 11
characteristic sequence, *see* CS
club, 10
collapsing function, 12
CS, 22

degree, 30
derivative, 11

Explicit Mathematics, 65

Finitist Arithmetic, 63
fixed point operator, 6
Frege structure, 64

Höwel, 68

join
    dependent, 6, 41, 62
    restricted, 47, 63
jump operator, 40, 42

linearity axiom, 19
Logic of Partial Terms, 3

Non-Finitist Arithmetic, 1, 17, 63
normal function, 10

operator, 24
    acceptable, 24
    Cantorian-closed, 24
operator-controlled derivation, 17
ordinal
    additively indecomposable, 11
    constructive, 52
    regular, 9
    strongly critical, 12

partial combinatory algebra, *see* PCA
PCA, 5, 60
progressive, 36

rank, 22
reflection principle
    local, 53
    uniform, 53
reflective closure
    ordinary, 57
    schematic, 57
replete, 67

SC, 15
schema, 55
    open-ended, 55
set theory, 50
Σ^ξ-sentence, 26
strongly critical components, *see* SC
substitution rule, 19, 34, 42, 55, 62

truth, 56
truth theory
    ordinary, 56
    ramified, 57
    self-applicable, 57
type theory, 49

unfolding, 2
    PCA-version, 5
    recursive definability, 61

Veblen function, 12
Veblen hierarchy, 12

www.ingramcontent.com/pod-product-compliance
Lightning Source LLC
Chambersburg PA
CBHW081829170526
45167CB00007B/2759